AF262638

TYLENOL AND AUTISM

Evidence, Scientific Blunders, and Medicine Gone Wrong

WILLIAM PARKER, PHD

Foreword by Dr. Randy Bollinger

Skyhorse Publishing

Visit our website at www.skyhorsepublishing.com.

10 9 8 7 6 5 4 3 2 1

Library of Congress Cataloging-in-Publication Data is available on file.

Cover design by Brian Peterson

Print ISBN: 978-1-64821-246-8
Ebook ISBN: 978-1-64821-247-5

Printed in the United States of America

For the babies of the world.

CONTENTS

FOREWORD

William Parker, PhD, has been labeled the "Tylenol-Autism Whisperer." He has an important message revealed in this book that has the potential to save many children and their families from the ravages of severe and profound autism spectrum disorder (ASD). I have known and worked with William Parker since 1993 when he joined the Duke University transplantation research group as our resident biochemist. I learned quickly that William has broad interests and deep knowledge in many areas of science beyond his formal training. While working in biochemistry/biophysics, he proved to be an innovative, unselfish collaborator who quickly gained the knowledge of transplantation immunology and gut microbiology needed for our projects.

Working at the interface between fields of science, William's questioning mind produced novel ideas and often led to new insights.

He questioned the conventional wisdom and, when that wisdom did not fit his experience and knowledge, he designed experiments to test the entrenched ideas.

His broad knowledge, innovative thinking, and creative experimental designs have led him to discover previously the function of the vermiform appendix and now acquire great confidence in the underlying trigger for all or most ASD. His conclusions regarding Tylenol and autism are based upon published experimental evidence from his and many other laboratories. The 341 references cited within this book are indicative of his careful, thorough approach to science. He realized that Tylenol (acetaminophen) had never been tested for toxicity during neurodevelopment before it was approved for pediatric use. He and others conducted critical experiments to test the conventional wisdom that Tylenol is safe. His conclusion that Tylenol is not safe for babies and young children is based on that evidence, not on opinions and not on conspiracy theories.

William's vast knowledge of this subject is remarkable. He was not the first to propose a link between Tylenol and ASD, but he had the intellectual curiosity to ask why and how they are connected. He challenged old assumptions and sought deeper explanations.

William is very, very curious, which leads him to new understandings. The medical community has not yet heard his message. Clinical practice has not changed. It is time for the public to hear Dr. Parker's well-documented, important conclusions. It is time for this book, written in language easily understood by the lay reader, to be available to the public. For the babies of the world, it is time for the "Whisperer" to shout.

Randy Bollinger, MD, PhD
Professor Emeritus
Duke University

INTRODUCTION

Don't take Tylenol. Don't give Tylenol to the baby after the baby's born.
—US President Donald Trump, September 22, 2025

Remember when, in the autumn of 2025, US President Donald Trump announced that acetaminophen, commonly known by the brand name Tylenol, can cause autism spectrum disorder? I was elated. My research team was elated. We had good reason to be. Like a raccoon set free from a trap, this information was finally out in the public domain, and nobody was ever going to put it back. It's easy to catch a raccoon in a trap once. But the animals are smart enough to remember the experience, and if you set one free, it can never be caught again using that same trap. From that day forward, any time a child regresses into autism, somebody—perhaps a parent or a friend of a parent—will wonder whether acetaminophen was present when the regression was triggered.

The immediate reaction of most people—including clinicians who had been prescribing acetaminophen to pregnant women, to women in labor, to babies, and to children for decades—was simple: "That's impossible." After all, scientists at Drexel University who had been looking at the connection between acetaminophen and autism had, a year earlier in 2024, laid the issue to rest, concluding that acetaminophen use during pregnancy was not connected with autism.[1] But then, in early 2025, just prior to the president's announcement, a report from Harvard University countered, saying that evidence of an association between acetaminophen use during pregnancy and autism was possibly significant.[2]

Chaos and confusion erupted. As a scientist working on the topic for more than a decade, I was interviewed dozens of times within days of the president's announcement, sometimes into the late evening hours. Other scientists weighed in on the topic, trying to explain evidence that my group had studied for years. Unfortunately, nobody got it right. The 2024 study from Drexel and the 2025 study from Harvard both missed the mark. The 2024 study from Drexel presented exquisite data—but completely misinterpreted that data. The study from Harvard presented much less convincing data, completely missed the bigger picture, but came closer to the truth for all the wrong reasons. Basically, Harvard was closer than Drexel to reality in terms of their conclusions, but only because their data were not very accurate. These two papers dominated the media coverage. The situation was surreal.

As you'll see in this book, the studies from Harvard and Drexel had very little to do with the actual connection between acetaminophen and autism because neither of those studies considered the vast majority of the available scientific evidence. Further, the conclusions of both studies were misguided due to sophomoric errors in the assumptions underlying their statistical analyses. I will explain all those errors in detail. But for now, imagine a group of very sincere people using a metal detector to find a woman's handbag lost in a field. But the handbag is plastic, not metal. A metal detector would be completely incapable of finding it— even though it's there. That's the situation.

In addition, both the Drexel and the Harvard study were focused on acetaminophen exposure during pregnancy. Available evidence, described in this book, tells us that *pregnancy is not the most critical period for the induction of autism.* Our very sincere group using metal detectors to find the woman's lost plastic handbag is also, for the most part, looking in the wrong field.

Chaos and confusion were followed by a growing consensus among scientists and clinicians that the president had missed the mark. Why did the administration of one of the world's most powerful nations make such a ridiculous official announcement?

I'm not saying here that this is a *true* consensus. As we will see, this is what is known as an "illusion of consensus." Illusions of consensus are created when one person promotes an idea or perspective, and others mindlessly repeat that same view.[3] Nobody is thinking critically, except the person who came up with the argument in the first place, who may, in fact, not be at all correct.

This book is all about the science underlying the president's announcement. The best available science at that time had enabled my research team to reach and publish the following conclusions:

1. Exposure of susceptible babies and children to acetaminophen causes many if not most cases of autism spectrum disorder.
2. The best explanation for all available evidence is that the vast majority of all cases of autism, perhaps more than 90 percent, is triggered by exposure of susceptible babies and children to acetaminophen.
3. Susceptibility is caused by a complex array of genetic and environmental factors that lead to *oxidative stress*, a condition in which the body cannot keep up with the chemical burden it faces.
4. The time of greatest risk for brain development is during and immediately after birth. Risk diminishes within weeks and dissipates almost completely by the time the child reaches six years of age. Some risk during pregnancy seems likely, but with less risk and with less certainty of risk than at birth.
5. Three rudimentary scientific errors—combined with subversion of the scientific process—have blinded most clinicians and scientists to the connection between acetaminophen and autism.

The president's assertion that women should avoid acetaminophen during pregnancy and not give it to their babies constitutes medical advice, not a scientific statement. Although ethical guidelines do not allow me or any other scientist to give medical advice, we are allowed to advise policymakers and medical professionals about what *their* policy or advice should be. Indeed, the foundation of modern medicine is science. If the conclusions above are correct, I believe that most people would agree that the US president gave good medical advice. It is my goal in writing this book to explain the evidence allowing us to draw those conclusions, providing a scientific basis for a much-needed course correction to medical practice.

At the time of this writing, few mainstream clinicians believe that acetaminophen is involved in the induction of autism. Paul Klotman at the Baylor College of Medicine provided an excellent example of the currently prevailing mindset. Dr. Klotman has been the president, CEO, and executive dean of the Baylor College of Medicine since 2010. Baylor currently has one of the best medical schools in the United States, and

Dr. Klotman regularly addresses students, faculty, and "friends of the school."

Dr. Klotman studies kidney damage related to infections with HIV, the virus that causes AIDS. Although Dr. Klotman has never professed to be an expert on neurodevelopment, drug interactions, or autism, he nevertheless went on record regarding the connection between acetaminophen and autism. On Halloween in 2025, Dr. Klotman addressed his audience:

> *So, this is to me almost like a manufactured controversy, and is amazing to me. And apparently the State of Texas is suing the manufacturers of Tylenol for whatever, for mislabeling, and this to me is honestly just silly. And, by the way, every professional society still recommends that for febrile illnesses during pregnancy that it is perfectly safe to take Tylenol. Most academic medical centers also make the same recommendation.[4]*

In this book we will review the evidence telling us, without reasonable doubt, that Dr. Klotman did not understand several critical issues. He didn't know how to interpret several landmark experiments, he wasn't aware of the vast majority of the relevant evidence, and he didn't even know when the developing brain is most at risk from acetaminophen exposure. But Dr. Klotman was correct about a couple of things. The State of Texas was suing the manufacturers of acetaminophen, and every professional society and most academic medical centers still claim acetaminophen is safe. This book will tell you why this situation constitutes a profound tragedy. Almost everybody is wrong, and they have been wrong for a long time.

This book explains why President Trump's advice makes sense, and Dr. Paul Klotman's does not.

The three major errors in the science regarding the connection between acetaminophen and autism will be described in plain language in this book. But scientific errors are not the only problem. Errors are an accepted part of science. They are unavoidable, but the discovery and correction of those errors is inherent to the scientific process. If the scientific process is allowed to progress, discovery happens and errors are weeded out.

The scientific process is the only way we have to unravel the mysteries of the physical universe—which includes the underlying causes of medical conditions. The scientific process works beautifully. The problems come when scientists think they know the answer and subvert the scientific process. As Max Planck, the renowned physicist, Nobel Prize winner, and originator of quantum theory, famously said:

> *A new scientific truth does not triumph by convincing its opponents and making them see the light, but rather because its opponents eventually die, and a new generation grows up that is familiar with it.*

This statement has frequently been summarized as "science moves forward one funeral at a time." This rather ghoulish-sounding assertion is known as "Planck's principle," and its impact is evident in the pages of this book. The legendary professor is saying here that the scientific process, which involves experimentation and analysis, is sometimes derailed by adherence to the dogma of the dominant scientists of the day. I prefer to think of it this way: Successful scientists typically move scientific understanding forward for years—and then hold it back for decades.

The subversion of science by senior scientists is not always evident. Most scientists know how much they can safely push against established thinking, and they know how much is *too* much. They simply never push against it enough to get bitten. This type of subversion is impossible to document. When scientists *do* push against established thinking, subversion can become evident, as you will see in this book. But incentives to cover up that subversion are generally strong. In a world where every bit of the funding a scientist receives is dependent on how *other* scientists judge their work, it does not pay to make enemies of those scientists by revealing their ethical lapses.

In this book, I will describe a number of ethical lapses in the scientific process related to the connection between acetaminophen and autism. I will give you a firsthand account of research projects being shut down using pressure that bears the hallmarks of retaliation, and I'll describe numerous breaches I observed in the accepted process scientists use to publish peer-reviewed scientific literature. Most importantly, we will see how the scientific evidence connecting acetaminophen and autism was obscured.

I understand that, when somebody says, "This is not sour grapes, I am just describing what transpired. I am not bitter at all," most people find it difficult to believe. But I'm really not just a bitter old man railing against the system. I will point out numerous signs of corruption in the system, but I have no reason for bitterness. First, I am absolutely not saying that anyone did anything immoral. I cannot and do not believe that anyone in the fields of science or medicine would knowingly sacrifice the lives of countless babies and their families, risking severe and profound autism spectrum disorder, for the sake of job security or career advancement.

Numerous historical figures have advocated the principle that, when a rule is unjust, that rule should be disobeyed. In other words, morality—good versus evil, if you will—trumps "ethical standards." The idea that unjust or harmful rules—or ethical standards—should be disobeyed is a universally accepted paradigm. I believe that the numerous breaches of scientific etiquette and standards that I will reveal in this book were performed by scientists and clinicians who believed they were taking morally correct actions. They believed they were protecting babies and children by blocking the conduct and publication of science that would lead to harm.

I can relate to a wonderful example of noble disobedience that occurred when a nature documentary crew filming the BBC Earth series *Dynasties* broke the cardinal rule of nature documentaries: observe, record, but do not interfere. In November 2018, they saved a group of emperor penguins in Antarctica, along with their chicks. The film crew needed only minutes to dig a few footholds in the ice using their ice axes, providing the penguins and their chicks a path to escape from a steep, slippery ravine where they had become trapped. Everyone was in favor. I would have done the same thing. Definitely.

Noble disobedience of the rules of nature documentaries is not the same as noble disobedience of the rules of science, however. Noble disobedience of the rules of science never accomplishes any goal other than to maintain wrong ideas and their supporters in positions of power and influence. Breaching the ethics of science will never be productive, and the underlying mental processes that lead to those actions are not rational. Scientists and clinicians are people, and like people in many other disciplines of life, they can be profoundly influenced by conflicts of interest, emotional compromise, and cognitive dissonance—all of which come into play here.

As a fellow human, I understand why other humans would choose to breach the rules of science. We have convinced ourselves that we are saving babies. I don't believe any of the attempts described in this book to impede research on the connection between acetaminophen and autism originated from evil intent. That is what I *believe*. What I *know* is that the scientific process works, but it can't work if we don't follow the process. Max Planck knew that. So, while I understand why people broke the rules, I know it can't work the way they hope it will.

In a further effort to convince the reader that I'm not simply blowing the whistle because of sour grapes and bitterness, I'll point out that I have not personally suffered from any of the events that I describe. My ever-patient wife and I did not lose our house and move into a basement owned by one of our parents—thanks in large part to generous donors to our nonprofit research and education foundation, WPLab. I have a wonderful community of supporting friends, and that community includes several scientific colleagues that still help me to this day. Even if I wanted to bring a lawsuit against someone who has attacked my career or my research, I would have no grounds for the suit. My wife and I have not been personally harmed, emotionally or financially. I may have missed out on some academic prestige, but who wants to be a prestigious scientist who has deduced what causes autism and does nothing about it? I prefer to be able to sleep at night.

I have no delusions about how science works in the United States. I am not, as far as I know, the victim of some horrible and unusual scientific conspiracy involving the pharmaceutical industry. The breaches in scientific ethics I'll describe in this book are typical in all fields of science that I have known. Max Planck made his observations in the field of theoretical nuclear physics, not autism. By the time I was a senior graduate student, before I even knew what autism is, the rules of modern science were apparent to me. If you anger the wrong person, your career is over. As somebody who stayed in the system for more than three decades, I knew the unwritten rules that underlie "success," and I had to deal with the reality of those unwritten rules to stay in the field of science. I can't complain about a system that benefited me for decades. But I can—and will—explain why people working in that system couldn't figure out the fact that a specific drug is causing permanent neurological injury in up to 3 percent of our population.

The point is that understanding breaches in scientific ethics is important in understanding how and why we missed the connection between acetaminophen and autism. It's possible that seeing how we missed the boat on this issue will help us avoid a similar event in the future. But I am less certain of that. History repeats itself. Human nature sees to that.

Now back to our story. The scientific method works, but only if we use it. When scientists start to break the rules, scientific progress stagnates, as Max Planck pointed out. In mid-2025, I encountered one of many examples of subversion of the scientific process when it comes to acetaminophen and autism. My research team and our collaborators had written a research paper describing two major scientific errors that have blinded many scientists and clinicians to the connection between acetaminophen and autism. The paper examines (a) when those errors appeared in the scientific literature, and (b) how prevalent those errors are today. We also provided some discussion from a historical perspective, pointing out that evidence available prior to 2009 should have been sufficient to change the way doctors practice medicine, dramatically reducing acetaminophen use before the age of six years. That research paper described the evidence in some detail and was eventually published in the *Journal of the Academy of Public Health*.[5]

The scientific peer-review process typically works this way: A scientist sends a research paper to a journal, which then assigns an editor. Importantly, when we send a paper to a journal, the paper is then "on hold" as far as we are concerned. We can do nothing with it until the editorial process is finished. If it takes an entire year, which is not uncommon, then we must wait for that year. Because of the value placed on the peer-review process, scientific work that has passed through the process carries much greater clout than work that has not. Although peer review is absolutely necessary to keep wacky and obviously wrong ideas out of the scientific literature, you'll see in this book that the peer-review process also creates barriers to the publication of innovative and game-changing work. Nevertheless, successful passage through peer review is necessary for any scientific work to gain acceptance by the scientific community and by the population in general. Therefore, no matter how long it takes, my research team submits all our work to peer review, and we pursue that process through to completion.

After the editor receives the paper, he or she obtains reviews from other scientists who are considered experts in the field or topic the paper covers. The editor then uses the reviews to decide whether the paper should be published. In addition to the options of accepting the paper for publication or rejecting it, the editor can and often does decide that the paper might be acceptable if the authors revised it in some way. The editor then sends the reviews to the authors and informs the authors of his or her decision. The authors are usually not informed of the reviewers' identities, unless a reviewer specifically asks to be identified. The editor normally informs the authors of the reason for whatever decision is made.[6]

We decided to submit this particular paper about the connection between acetaminophen and autism to the *Journal of Xenobiotics*. After a month, we received two anonymous reviews and a rejection decision by the editor, who in this case was also anonymous. The full reviews, along with my responses to those reviews and the editor's final decision, can be found in Appendix A. One reviewer was very positive, describing our work as a "well written systematic review" and suggesting additional detail be added in several places. Such constructive reviews are an important and vital part of the scientific process that improves the quality of the published work. However, the other review was vitriolic and incredulous. The reviewer asserted that the conclusions of the manuscript were "outrageous, overstated, illogical, and unnecessarily inflammatory." The primary scientific argument provided by the anonymous, negative reviewer was as follows:

> *Nearly every child is exposed to acetaminophen. Thus, saying that acetaminophen exposure causes autism in susceptible children is the logical equivalent to saying that breathing air causes autism in susceptible children.*

This is a two-part argument, neither of which has any validity. The first part of the argument is that, since everybody is doing it, it must be okay. This is a favorite argument of children who want their parents to allow them to do what their friends are doing, of course, and is patently invalid. The second part of the argument is that, because it is so commonly used, then it's impossible that acetaminophen adversely affects some people and not others. That argument is also not valid in any way. Numerous chemicals and drugs affect some people more than others, both positively and negatively. In many cases, the reason some people are more sensitive or susceptible than others is well known. Later in this book, we will consider the widely used drug codeine as a classic example. The expert reviewer summed up with an absurd and unsupportable statement comparing the safety profile of acetaminophen with the safety profile of the air we breathe.

The antagonistic, anonymous reviewer of our paper also accused us of "raging against straw enemies" (Appendix A). The *straw enemy fallacy*, more commonly known as the *straw man fallacy*, is defined as diverting attention away from one issue by raising some other issue that is completely irrelevant and may even lack any basis in reality. In reality, the paper—which is now available for anyone to view[7]—quantitatively and systematically documents several errors in the scientific literature related to the connection between acetaminophen and autism. Since the errors are quantifiable, and indeed were pervasive in the scientific literature, they are, by definition, *not* straw enemies. They are real. Ironically, therefore, and again by definition, our antagonistic reviewer raises an unreal "straw enemy" in order to divert attention away from the real issue—quantified misinterpretation and mishandling of evidence related to the connection between acetaminophen and autism.

Another argument in the antagonist review, made without any valid or rational justification, is that our conclusions are "illogical and untrue." This is a logical fallacy known as an *argument from authority*. "Because I said so" is often helpful for parents when they don't think their child will grasp an underlying rationale, but that argument has no place in science. Further, the fact that the reviewer did not believe our conclusions, despite the fact that we provided ample evidence supporting them, is another fallacy, the *invincible ignorance fallacy*. You'll see examples of that fallacy throughout this book.

We pointed out to the editor of the *Journal of Xenobiotics* that the reviewer appeared to be biased, since they did not provide any valid arguments refuting any of the evidence we presented. In fact, the reviewer did not address even a single point of the evidence we presented (Appendix A). But our protest was ignored. After about two weeks, Veralyn Kong, the assistant editor, informed us that an anonymous senior editor was standing by the decision to reject the paper based on the—obviously invalid— review. The journal's senior editor did not follow the accepted scientific process, which is to find another reviewer after the overtly invalid review was received. This ethical breach needlessly delayed publication immediately prior to President Trump's announcement described earlier. Had the paper been published in a timely fashion in a peer-reviewed journal indexed in major medical databases, it could have addressed much of the controversy that soon surrounded the president's announcement. My research team wrote that paper in collaboration with Professor Paul Corrigan, an award-winning poet who teaches persuasive writing to university students. With Paul's coaching, we tried very hard to make the paper readable for the nonexpert, and I would encourage anyone interested to take a look at it.[8]

The *Journal of Xenobiotics* is managed by the publisher MDPI and is indexed in major medical databases. Some might suggest after hearing our story that journals published by MDPI are, in general, not as reliable as journals from some other publishers. However, you'll see in this book numerous examples of subversion of the science related to acetaminophen and autism, and some of those ethical lapses happened at some of the most prominent journals in the world. You'll also see an example of brilliant work by editors working with MDPI who are not afraid to attach their names to the papers they usher through the peer-review process. In my experience, individuals make a difference, sometimes in very wonderful and sometimes in very harmful ways. It's not the prestige of the journal that drives science forward or backward; it's the people handling the science.

This book will describe the scientific evidence connecting acetaminophen and autism, point out some of the ethical lapses subverting that science, and explain the technical errors in scientific exploration that have tragically affected our understanding. The reader will see how it is possible that elite medical professionals—such as Paul Klotman at the Baylor College of Medicine—could be completely ignorant of one of the most important medical issues of our time. The purpose of this book is to reveal what the experts missed and how they missed it. After you've read this book, you'll be able to answer the question, *Did President Trump give good medical advice?*

By way of warning, much of this book is going to read like a dark comedy. I am keenly aware that this topic affects many people very deeply. At the same time, to be effective, I cannot take the events described in this book personally. When I face a decision by a colleague or an editor that shuts down six months of effort to inform the public about avoiding the needless induction of autism in babies and children, I can't afford to lie awake at night stewing in anger. Too much is at stake for me to take any of this personally. Emotion blocks the ability of humans to think logically and strategically, and I am not immune to that effect. In the business of science, logic and strategic maneuvering are critical.

Throughout this book, I'll describe numerous events involving scientists who have made profound mistakes, sometimes allowing personal connections and emotions to impede their judgment. You'll see me present events with a matter-of-fact perspective, without railing at the potential damage that specific decisions and actions may have caused. I'll sometimes use easy-to-understand, even silly examples and analogies to help explain scientific principles. It is not my intent to belittle the importance of the subject material, but you'll see from my writing that this story is not deeply personal for me. It's just science—and for me, for now, it has to be that way if I am to be effective. When this is finally all said and done, then I will have the freedom, or perhaps the obligation, to embrace the personal impact of this work on the lives of so many.

CHAPTER 1

THE TYLENOL-AUTISM WHISPERER

I was trained as a scientist in a classic fashion. Through the efforts of gifted and caring teachers, I fell in love with science in middle and high school. I was always excited by my science projects in the rural school system of White Hall, Arkansas, but those projects didn't always do well in competition. In my senior year, I designed an easy-to-assemble type of temporary greenhouse. I monitored the growth of plants and the temperature of the system during the winter months to determine how much extra growing time the greenhouse yielded. But my project was not the best. One of my classmates was particularly gifted in math and had solved a classic math puzzle in a way that had never before been accomplished. Based on my subsequent involvement with science competitions as a mentor, I now realize that my classmate's work could have achieved national recognition. At that time, at least I knew he should have won the competition.

In the science fair judging, a classic volcano demonstration was declared the winner. The classic volcano demonstration has several variations, but in general it's a hollow, volcano-looking sculpture made of clay or some other easily formed material, then loaded with baking soda and vinegar. When the vinegar and baking soda come in contact, they react to produce foam, which pours out the top of the volcano-looking sculpture, creating a simulated lava flow. The drama makes it a crowd-pleaser, and the simplicity of the demonstration makes it popular in elementary schools. Fortunately, my loss as a high school senior to an elementary school–level volcano did not deter me from science. But the event demonstrated to me at an early age that science judges are not always competent, and I believe that was a helpful lesson.

I earned a bachelor of science degree in biology and a bachelor of arts degree in chemistry at the University of Arkansas at Little Rock in 1987, conducting undergraduate research with Robert Steinmeier, who had trained at the University of Nebraska in Lincoln with Larry Parkhurst, a renowned biophysicist. Professor Steinmeier had a small lab, with no federal funding and only a couple of undergraduate students. He didn't consider publication of scientific papers a priority and focused on training. I learned from Professor Steinmeier what sorts of wonderful things could be done with skill, great ideas, and a shoestring budget.

I went on to earn a PhD in chemistry at the University of Nebraska, working in the laboratory of Pill-Soon Song, another renowned biophysicist and editor of the *Journal of Photobiology and Photochemistry*, the official journal of the American Society for Photobiology. The work in Professor Song's lab, funded by the National Institutes of Health and the US Department of Defense (now also known as the US Department of War), provided me with excellent training. By the time I graduated from the PhD program, I was a competent protein biochemist, having published several papers in plant biochemistry and biophysics. But, perhaps just as importantly, I had experience running "projects on the side" with work that wasn't funded by a grant from any agency, but which could be done cheaply. My first side project involved a model of protein folding.[9]

Also in graduate school, I got a taste for the wonderful world of collaboration. I had the privilege of collaborating with Eefei Chen in David Kliger's lab at the University of California, Santa Cruz. Using a type of ultrafast machine that can see protein shapes, we held the world's record for the fastest observation of protein structure.[10] It turned out that our protein was slow moving, and we didn't actually need the incredible speed in our system to see the protein move, but we had the world record, nonetheless. Essentially, we had an amazingly fast camera—capable of watching a speeding bullet—trained on a turtle. At least we could prove that the turtle didn't move fast. Eefei was the first of several collaborators that I count as friends to this day. A wonderful collaborator is as good as gold in academic science.

After graduation with a PhD in December1992, I began working in the laboratory of John Stezowski. John's specialty was shooting X-ray beams through crystals and looking at how the beams scatter. The pattern of scattering allowed John to determine the three-dimensional structure of whatever was in the crystal. My work in John's lab was fun and led to a publication in the field of X-ray crystallography,[11] but the main reason I wanted to work with him was because he was a great help to me on my side project involving protein folding.[12] Susanne, my ever-patient wife since 1989, supported us as I worked on that project during the spring semester of 1993 without a salary. I taught biochemistry at the University of Nebraska that summer, and then we moved to Duke University for a two-year postdoctoral fellowship, where we stayed for twenty-seven years and eleven months. I should clarify that most people would probably not have considered Susanne to be "ever-patient" back in 1992, but she certainly earned that badge by the time I left Duke in 2021.

Duke University Medical Center held the promise of a postdoctoral position with Jeffrey Platt, a pediatric nephrologist turned scientist in the Department of Surgery. Jeff was, I believe, the most creative scientist I have ever known. He had incredible imagination and applied it to science in beautiful ways. But Jeff didn't know the nuts and bolts of running a biochemistry experiment, and he needed a biochemist to execute his ideas. I was that biochemist. Jeff's lab was wonderfully funded by the NIH and by private companies, and we applied that funding to the field of *xenotransplantation*, or cross-species transplantation. During that time, I was privileged to be involved in some of the first transplants using pigs with human genes.[13] The idea was to genetically modify the pigs in a way that would make their organs compatible with humans for transplantation as a treatment for organ failure. The technology at that time was in its infancy, and Herculean efforts were required to produce those pigs. It was from Jeff and my close association with other wonderful scientists at Duke that I learned immunology and a great deal about the field of medicine.

It was also at Duke where my penchant for shoestring-budget projects on the side blossomed. We always had some "discretionary funding," not earmarked for any particular project. That funding was generally provided by the department, although the university also chipped in.

In addition, my ever-patient wife allowed me to spend considerable amounts of our personal income on my laboratory work.

I remember the moment that—as I was looking at some unexpected results from an experiment I had run for Jeff involving xenotransplantation—it occurred to me that the immune system *supports* the growth of intestinal bacteria.[14] That moment was exciting. I immediately realized that my conclusion was (a) apparently correct based on everything I knew, and (b) made much more sense than conventional thinking in the field of immunology at the time. The prevailing view at that time in immunology was that the immune system was constantly fighting off our own intestinal bacteria. The more reasonable way of looking at the immune system involved its *cooperation* with intestinal bacteria—which made sense when taking a step back and looking at the big picture. I had come to my conclusion based on the biochemistry of the system, but considerations from the fields of ecology and evolutionary biology pointed to the same conclusion. That new understanding of the gut microbiota opened a flood of new ideas, and after Jeff moved to the Mayo Clinic in Rochester, Minnesota, I began to work with Randy Bollinger, a surgeon and immunologist, on those new ideas at Duke.

My initial work on the immune system cooperating with bacteria eventually led Randy and me to discover the function of the *human vermiform appendix* as a kind of "safe house" for beneficial bacteria in 2007.[15] Work by other scientists continued in that field, eventually resulting in the discovery that the presence of an appendix is associated with protection of primates against severe diarrheal disease.[16] Such protection can be explained and predicted if the appendix serves as a safe house for beneficial bacteria.

This side project describing the immune system's relationship with gut bacteria gave me my first taste of the difficulty in overcoming entrenched scientific ideas. Despite multiple attempts, Randy and I could never get the project funded through the National Institutes of Health. I remember one anonymous reviewer who stated derisively, "The authors appear to have never heard of mucus." The idea behind the comment was that it's obvious to everyone that the purpose of mucus in the gut is to keep bacteria *out*. Indeed, that was the thinking at the time. But we knew that gut bacteria embrace mucus, so to speak, sticking to it using specialized molecules called *receptors*. If the mucus were trying to keep the bacteria away, why did the bacteria stick to it? Why didn't the bacteria simply stop making those receptors so they could evade the mucus and go around it or through it? After that review, it took us only a few weeks to do the experiments showing that the mucus was part of our immune system's welcome service and maintenance program for our own beneficial bacteria.[17] In a nutshell, our body's immune system helps gut bacteria form living films, called *biofilms*, which help keep the bacteria inside our gut.

My side projects have been exceedingly rewarding from an intellectual perspective. One of my high school interns, Ryan Lee, got to shake hands with President Barack Obama because his work on protein folding[18] was recognized in the Intel Science Competition of 2011. My lab was the first to develop an artificial model of the human gut containing living human gut cells and living gut bacteria,[19] and we were the first to observe living bacterial biofilms in the healthy human gut.[20] My lab's description of the function of the human appendix answered the question, *What is the function of the human appendix?* The question was first considered by Leonardo da Vinci more than four hundred years earlier, and was again considered by Charles Darwin, the legendary biologist.

We were the first to compare immune systems between wild and laboratory animals,[21] gaining profound insights into what "being clean" in modern society can mean for immune function. We also published the first systematic surveys of the beneficial effects of intestinal worms,[22] finding that certain intestinal worms can alleviate common neurological illnesses such as major depressive and anxiety disorders.[23] Because of my work on wild rats and intestinal worms—and contrary to the expectations of almost all other experts—I correctly predicted[24] that intestinal worms would protect people from adverse clinical outcomes resulting from COVID-19 infection.[25] In another project, we discovered genetic sequences of hundreds of new bacteria never before observed, finding that tens of millions of years ago, a ratlike species established a partnership with bacteria from a cowlike species.[26] Most recently, working with Kateřina Jirků and a wonderful team of scientists at the Czech Academy of Sciences, I co-discovered the fact that intestinal tapeworms are capable of a kind of hibernation-like state called *aestivation*.[27] They can go dormant and survive in the gut for significant periods of time if their host isn't eating—when the host is in hibernation, for example.

My faculty position at Duke was funded to some extent by administrative duties, but mostly through well-funded clinicians working in the field of transplantation. My own projects, unrelated for the most part to the field of transplantation, were never funded by any major federal grant and remained part-time endeavors. My personal projects were, essentially, *not* fundable. For example, federal funding is difficult to obtain for finding that a benign (harmless) intestinal worm alleviates depression, even if the biology makes perfect sense.[28] And it's not easy to get federal funding for developing clinical trials aimed at treating depression with a benign intestinal worm.[29] Federal funding comes from being on the cutting edge of the molecular underpinnings of depression, which can potentially lead to the next blockbuster drug. I never gave up trying to get federal funding, but I never came close. None of my applications made it past the first round of review. Fortunately, my clinician-scientist colleagues seemed pleased with the scientific support I provided for their well-funded work in transplantation science, and I eagerly applied my ever-increasing scientific knowledge base to help them with their laboratories. I appreciated the privilege of working at the interface of medicine and science and was indeed excited to be a part of the team of clinicians and scientists. I felt needed and appreciated and published dozens of papers with my colleagues in the field of transplantation.

I did well enough in terms of scientific publications. I have about 150 of those in the peer-reviewed scientific literature, which falls in the mediocre range for a senior scientist from a major research university. That's not even close to the top-performing scientists. To give you an idea of my record, we can look at the "h-index," one way to measure productivity with publications. The h-index is a combination of how many publications someone has and how often their publications are cited. An h-index of 100, for example, means that somebody has 100 papers that have been cited at least 100 times. The h-index isn't perfect. For example, if a scientist has only one paper, but it has been cited one billion times, then their h-index is still only one. At the time of this writing, I have an h-index on Google Scholar of 53, which is good, but definitely not great, for someone who has been publishing research for more than thirty years. By comparison, Kent Weinhold, my immediate supervisor when I was in the Department of Surgery at Duke, has an h-index on Google Scholar of 74 at the present time. That's very good. Kim Lyerly, a renowned clinician scientist who does cancer research at Duke, has an h-index of 89 on Google Scholar. That's impressive.

When I retired from Duke, I was an associate professor without tenure. While my h-index is mediocre for a senior research scientist, it is absurdly high for an associate professor without tenure. Normally professors move up the ranks from associate to full professor when their h-index is in the mid-20s. But without significant research funding, I was never going to move up the faculty ranks, and I understood that fact. I made the decisions that put me where I was, and I was happy. The graph in Figure 1.1 sums up my scientific career in a nutshell.

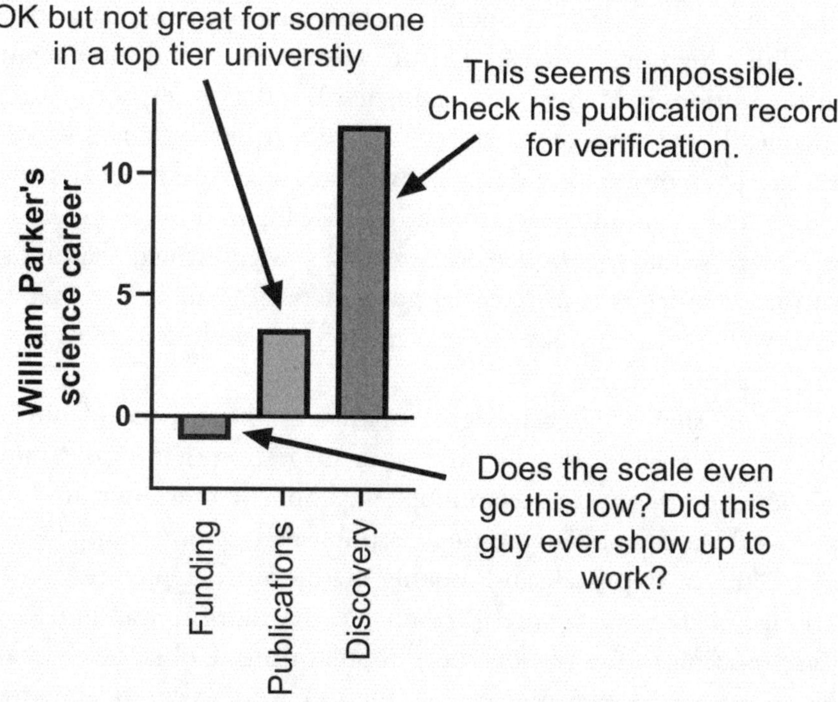

Figure 1.1: A synopsis of William Parker's performance as a scientist. The axis on the left, called the *Y-axis*, is labeled from 0 to 10, but the numbers don't correspond to anything more than 0 = very bad, 5 = average, and 10 is amazing. This is what scientists call a *schematic diagram*. It gives people an idea of reality, but it's not based on exact measurements collected in a real experiment.

My side projects continued. Because of my lab's discovery of biofilms in the healthy human gut in 2003,[30] I was contacted in 2007 by Laura Bono, who worked with an organization called SafeMinds, which was interested in the gut function of children with autism. It was recognized at that time that children with autism tend to have more problems with gut function than neurotypical children. With that in mind, I began lecturing about normal gut function at some meetings for parents and clinicians interested in autism and got to know many wonderful people in that community. As a scientist interested in the underlying causes of disease, autism was intriguing to me. It was clearly related to chronic inflammation in modern society, which was something I appreciated based on my work in immunology, and especially my work comparing wild and laboratory rats. Allergic conditions, autoimmune diseases, and many neuropsychiatric conditions had some properties in common and share common underpinnings in high-income, modern societies.[31] Perhaps autism was simply another manifestation of the same issues that led to a high prevalence of allergic, autoimmune, and some mental health conditions in high-income countries? But the timing of the autism pandemic did not seem to match. . . . Why was it so recent, and why the broad spectrum of the condition? And why was it so hard to find a particular underlying immune reaction that was causing it? Something did not seem to fit. Research conducted after 2017 has confirmed that autism spectrum disorder is connected to gut function,[32] but it was Stephen Schultz who provided a possible explanation that addressed most of my questions about autism.

In 2008, shortly after I began looking at the science of autism, Stephen Schultz published the first work suggesting that autism might be an adverse reaction to acetaminophen.[33] Shortly thereafter, in 2009, Peter Good, an independent researcher, published a more comprehensive paper, taking a step back and looking at the broader picture.[34] Good concluded that it was entirely reasonable that autism was indeed an adverse reaction to the world's most popular drug. Unlike Peter Good, I was constrained by training and conditioning to accept an unwritten academic rule: without years of mentorship and thousands of hours of study, I lacked the credentials to publish anything on the cause of autism. Indeed, many scientists have become renowned in one field and then blundered into another field, making sophomoric errors. I was not the blundering type. But Schultz and Good both made sense to me.

In 2010, I told a family member who was watching his second child regress into autism that I thought that acetaminophen was likely instrumental in the regression. The parents of the child immediately discontinued acetaminophen administration, and the child ceased to progress down the path of regression that her older sibling had traveled only a few years before. But I remained bound by my training and the standards of science as I understood them. I followed the literature closely but stuck to my fields of expertise, which were, at that time, gut function and transplantation. I was sure that some qualified expert would rapidly and incisively deal with this issue of acetaminophen's connection with autism.

By 2013, Bill Shaw—another independent researcher unconstrained by internal reservations about publishing something outside of his field of expertise—had followed in Peter Good's footsteps, publishing yet another paper that looked at the big picture and concluded that autism spectrum disorder is likely an adverse reaction to acetaminophen.[35] By 2015, I was still sticking to my fields of expertise for the most part, working with a wonderful team of scientists putting together a paper on modifying gut function in children with autism.[36] But I had also worked with the renowned neuroscientist Staci Bilbo, writing about the possibility that acetaminophen was involved in the induction of autism.[37] However, that paper was mostly focused on the general idea that chronic immune dysfunction associated with Western society was connected with autism.[38] Nobody reading the title, abstract, or introduction of that paper would have known that it had anything to do with acetaminophen. Information about acetaminophen was discussed extensively in the manuscript, but that discussion didn't start until the third page.

After the 2015 paper was published, I was surprised that nothing happened to change clinical practice. Duke University professors had pointed a finger at acetaminophen in the induction of autism, and nobody seemed to care. My previous work on the appendix, wild rats, and even the benefits of harmless intestinal worms had received considerable media attention. I realized I was naive to expect clinical practice to change based on some statements we had buried in a scientific paper and decided that we needed to step up our game a bit. My lab's first paper dealing strictly with the connection between acetaminophen and autism was published in early 2017.[39] Professor Bilbo, the renowned neuroscientist, was a coauthor, and Martha Herbert, a pediatric neurologist and professor at Harvard, was part of our team. That paper was the first to connect the multitude of genetic, epigenetic, and environmental risk factors for autism with acetaminophen in a logical fashion that fit all the available evidence. At that time, we had fourteen lines of evidence, less than half of what we have today, pointing to a causal relationship between acetaminophen and autism. Further, many of those fourteen lines of evidence are stronger today than they were in 2017. But even then, in 2017, the science made sense. The response was not what I expected.

Following that 2017 publication, the Department of Surgery at Duke University Medical Center implemented a series of actions that are often used in retaliation:

- My administrative duties were removed without notification, putting my job at risk due to loss of salary support. In academia, salary support comes from teaching, administrative work, and grant funding.

- I was assigned tasks that had no academic value. Examples include helping other faculty organize their schedules and creating a mural honoring a deceased immunologist.

- I was not allowed to use my discretionary funding for my research. My ever-patient wife, Susanne, intervened, borrowing against our house to support the work. Eventually, the Surgery Department refused even to accept donations to support my salary.

- I was blocked from using media relations, donor outreach, and logistics services. I literally could not order supplies for my laboratory.

- I was formally accused of abusing employees, without warning and without being told who was abused, how they were abused, or what I should do differently. I was told radically different things in person than I was told in writing. This administrative maneuver violated several personnel management policies described in Duke's management training program. Ironically, adult children of other faculty worked in my laboratory, and my employees had on more than one occasion asked if their adult children could work in my laboratory. That is not the sort of thing anyone does if their boss is abusive. I consider myself a "people person," and my faculty colleagues expressed shock when they learned of the formal accusations.

If those actions were retaliatory, they are legal in North Carolina, where Duke is located. Retaliation is illegal in North Carolina only when it involves whistleblowing about illegal activities or involves issues related to discrimination (for example, racism, sexism, or ageism). Nevertheless, friends offered to order my laboratory supplies for me, and we found ways to work around the other problems I encountered, continuing my work on acetaminophen. We found that doses of acetaminophen proportionally lower than those administered to babies and children caused permanent changes in social behavior in laboratory rats.[40] Although I had obtained adequate private funding to support my work and my salary, my laboratory was shut down in June 2021 with the only reason given that it was not in the "strategic interest" of the Department of Surgery to keep my lab open. At the time, I had a signed three-year teaching contract with Duke. But without a lab to conduct research, there was no point in teaching, and I retired from Duke, abandoning the teaching contract to set up a private, nonprofit laboratory and continue my scientific work.

One of the sacrifices I made to conduct my on-the-side projects was that I had to accept that federal funding was never going to be likely. Except in rare cases, when scientists fail to get past the first round of grant review after repeated attempts, it's usually necessary to consider the project unfundable and scrap it. I had a stable (I thought) job and enough discretionary money to run my side projects, so giving up those side projects to pursue other less intriguing projects that might attract federal funding made no sense to me. The drawback was that I was an associate professor without tenure and without the prospect of tenure. Tenure would have provided me with some security, preventing administration from shutting down my laboratory without valid reason. But without federal funding, which is a key indicator of success in modern science, there is no way to get tenure.

It was in 2022, a year after my laboratory was shut down, that Jennifer Margulis, an investigative reporter, informed me that A. Eugene Washington, the president and CEO of Duke University, was on the board of directors for the manufacturer of Tylenol—the major brand of acetaminophen in the United States. Duke had a profound conflict of interest in shutting down my laboratory. For the first time, I could put my finger on a reasonable explanation for everything that had gone wrong for me at Duke. Ironically, it wasn't until I had begun work at the University of North Carolina, Chapel Hill, and was taking exactly the same management training that was mandated every year at Duke, that I recognized the classic pattern of retaliatory activity. Duke administrators had applied literally every example of retaliatory activity provided in the university training to *me* following the publication of our 2017 paper linking acetaminophen with autism.

It's important for me to acknowledge that I do not know who was pulling the strings that impeded my research on acetaminophen at Duke. I never had any interactions with Eugene Washington. As far as I can recall, we never met or even corresponded by email. I do not know that he had anything to do with the efforts that hampered—and eventually halted—my research at Duke. While Washington had a clear conflict of interest, his position with the drug manufacturer was not the only suspicious connection involving Duke. For example, Mark McClellan, the director for Duke's Center for Health Policy, an influential position within the university, was also on the board of directors for the manufacturer of Tylenol. And the problems my research faced at Duke could literally have been the result of "pillow talk." Mary Klotman was the dean of the Duke University School of Medicine starting in 2017. That made her the boss (dean of school) of the boss (chairman of department) of my boss (vice chair). It turns out that Mary Klotman has been married to Paul Klotman since 1981. You may recall from the introduction that Paul Klotman views the idea that acetaminophen is involved in the induction of autism as "silly."

Though I don't know who pulled the strings at the top of the ladder, it was Allan Kirk, chair of the Department of Surgery, who *directly* impeded my research on acetaminophen. And it was Allan, along with Kent Weinhold, vice chair of the Department of Surgery, who signed the letter that shut down my laboratory. As I left Duke, Allan and Kent went out of their way to block my access to academia. Even though I had funding remaining with the Department of Surgery, the department denied me an adjunct appointment, which would have cost them nothing but would have allowed me to keep my Duke email address and finish up my work at Duke. Again, I had help from friends and was able to work around the hurdles imposed by Duke as I retired from the university.

I would probably never have left Duke while my laboratory was running, regardless of the amount of abuse I received. I knew the prestige of Duke lent credibility to my work, and it would have been selfish to abandon that credibility for personal comfort. More importantly, I was not marketable to another university because I had no funding in my own name, and no university would hire senior faculty without a long history of federal funding. Thus, it was Duke or nothing—at least as far as I could see.

When I left Duke University for the last time, my car loaded with my personal office supplies and most of my research notebooks from my side projects, I stopped at the first garbage barrel I found, took the shoes off my feet, and thew them in. It was a symbolic gesture that helped me remember to look forward. At that time, one child was being diagnosed with autism about every five to six minutes in the United States alone. It was time to get to work without the dense baggage imposed by Duke's administration.

Being free of Duke and setting up my own nonprofit research organization had benefits I could never have foreseen. As head of a private nonprofit organization, I was able to collaborate with other universities that were much better equipped than Duke to run the experiments on rat behavior that I was interested in. I was extremely blessed to be accepted as a collaborator and visiting scholar by Kate Reissner, a neuroscientist at the University of North Carolina in the Department of Psychology and Neuroscience. Kate began her scientific career working with me in the field of transplantation science a quarter century earlier[41] and had long surpassed me in terms of faculty advancement and rank. I was also blessed to collaborate with Lauren Williamson, a professor at Northern Kentucky University, who had also, not surprisingly, surpassed me in terms of faculty advancement and rank. Lauren and I had worked together while she trained with the renowned neuroscientist Staci Bilbo,[42] who has served as director for Harvard's Laurie Center for Autism. Lauren's expertise in the field of neuroscience complemented the expertise we had in Kate's lab at the University of North Carolina, and the arrangement allowed us to independently confirm experimental results. The fact that Lauren and Kate both agreed to work with me on the connection between acetaminophen and autism was exciting.

Retiring from Duke did not affect my strong working relationship with J. P. Jones and Zacharoula Konsoula, a clinical pharmacologist and toxicologist, respectively. Their knowledge and training continued to be vital to our work. I also continued work with Randy Bollinger, a transplant surgeon and immunologist who had retired from clinical duties a few years before I retired. In addition to these benefits, my nonprofit allowed me to work with marketing and communication experts in ways that would never have been possible within the confines of academia.

After I left Duke, our amazing team of wonderful collaborators proceeded to publish several papers examining the broad spectrum of evidence connecting acetaminophen with autism.[43] At the time of this writing, the first and last authors of our most recent paper,[44] still under review, are the same scientists who discovered the function of the human appendix—Randy and me. One of our papers,[45] published in the flagship journal of the Korean Pediatric Society, has been viewed more than 135,000 times. Progress was quicker without the baggage from Duke.

It is not possible to induce autism in 3 percent of the entire population without leaving an obvious trail behind, and our goal was to keep probing that evidence and bringing it to the attention of the public. Peter Good and Bill Shaw also continued to publish great work on the topic,[46] with an emphasis on environmental factors that could interact with acetaminophen. And, as will be explained in this book, dozens of other investigators were working tirelessly on the issue as well, publishing both work in laboratory animal models and analytical work examining health-care data. Tragically, some of that work was crippled by scientific errors, another topic that we probed in our scientific work and that will be explained in this book.

The 2024 presidential election changed the landscape of my work significantly. My connections with the community of parents and doctors focused on autism science ensured that I knew people associated with Robert F. Kennedy Jr. (RFK Jr.), who had also been working with parents of children with autism for years. After Donald Trump formed an alliance with RFK Jr. and won the 2024 US presidential election, I was confident that my work and the work of others on acetaminophen and autism would come to light. The evidence connecting acetaminophen with autism was more than strong, as will be explained in this book. In addition, the scientific errors that misled investigators who concluded that acetaminophen is not involved in the induction of autism were obvious. The bottom line was that the evidence was incredibly strong, no valid objections remained, and somebody who was motivated to bring the connection between acetaminophen and autism to light was now in a position of power to do it. Everything was set to go the day after the 2024 US presidential election. When Susanne and I woke up on the morning of November 5, 2024, we knew it was only a matter of time before the world would know of the connection between acetaminophen and autism.

News that I had discussed the acetaminophen/autism connection with RFK Jr., secretary of Health and Human Services, somehow leaked to the press in 2025, and I was contacted by a reporter for *The Atlantic*. At that point, I knew that, if *The Atlantic* didn't abandon the story, the raccoon would be out of the trap, never to return. I spent over two hours reviewing the evidence with the reporter, and was elated that he appeared willing to listen. The article was published on September 9, 2025, at 5:00 p.m. ET, and painted a picture of me as a fringe scientist who worked in relative isolation. It was not what I had expected. I had been warned that my new "friends in the media" might not simply focus on the science, as I had experienced in many dozens of previous interviews involving my side projects of the past. My friends were correct.

I had corresponded with *The Atlantic*'s reporter using my university email address, and he was aware that I had collaborated with faculty from half a dozen universities since leaving Duke, so to label me as operating "on the fringes of academia" was questionable at best. He also knew we had published our papers in several notable scientific journals. This is not to disparage scientists who *do* work on the fringe. Many great leaps in science are made by such people, but I don't think I ever earned that label. My work is very interdisciplinary, and strongly dependent on collaboration. On the other hand, I may have earned the label of "Tylenol-Autism Whisperer," used in *The Atlantic*'s title, as long as we keep in mind that (a) many other scientists are working on this issue, and (b) I work with a wonderful team of neuroscientists, industry professionals, and brilliant students who volunteer their time.

The Atlantic's article permanently changed something—it brought the public's attention to an issue that, we have concluded, is one of the most important in medicine today. The pathetic scientific evidence presented by *The Atlantic* is not reconcilable with the overwhelming evidence their reporter was given, and that is indeed a tragic and unforgivable disservice to their readers. But the article did serve the purpose of getting people to look at the evidence, the same evidence that I'm happy to present in this book.

When examining the evidence for yourself, please keep in mind the scientific reasoning we applied when drawing our conclusions. As with our previous work on the interactions between the gut and its bacteria, no single piece of evidence allowed us to draw any conclusions. Some lines of evidence are indeed very compelling, but they are not conclusive when considered alone. It is the weight of all evidence, taken together, which in this case allows conclusions to be drawn. Importantly, that evidence includes evaluation of alternative explanations. I address these alternative explanations, as well as all objections to our conclusions that we have seen in the scientific literature, and some that we have seen only on social media or in the press, in this book. I will also explain key errors in data analysis and breaches in the scientific process have tragically prevented most scientists and clinicians from recognizing the connection between acetaminophen and autism.

CHAPTER 2

THE HISTORY OF ACETAMINOPHEN AND ITS CONNECTION TO AUTISM

A century of exposure to acetaminophen before anybody even thinks to ask the safety question. (And then nobody apparently wanted to know the answer.)

Many narratives describe a "cause" of autism. Often those narratives start with a statement along the lines of "autism has increased dramatically since" whatever the author thinks causes autism was introduced into the population. I could do that, but then I would lose many of my most critical readers, and I said I would cover all the objections to our conclusions. I will examine the idea of an increase in the prevalence of autism from a perspective I haven't yet seen in the scientific literature. Instead of trying to draw a conclusion based on one bit of information, diverse lines of evidence will be considered.

It has been argued that, since the first descriptions of autism predate the use of acetaminophen, then acetaminophen *can't* be a critical part of causing autism. In a down-to-earth example, one "expert" explained on social media that, since her father has autism, and since he was born before 1954 when acetaminophen was introduced, acetaminophen can't cause autism. In this case, the expert was referring to the year 1954, when acetaminophen was brought to market. However, the reality of the situation is that, to *know* her father had avoided acetaminophen as a child, he would need to have been born prior to 1886—which would make him about 140 years old when the expert provided her comment on social media. The expert did not reveal the actual age of her father but did suggest he was still alive (she said, "my father has autism," not "my father had autism").

Acetanilide, which began to be used in 1886, is converted by the body into acetaminophen. *Phenacetin*, also converted by the body into acetaminophen, was introduced the following year, in 1887. We know that at least a few children were exposed to phenacetin during those early days,[47] but for the most part, nobody has tracked pediatric use of acetaminophen or the drugs that are converted into acetaminophen by the human body.[48]

The first clinical description of autism as we know it—complete with repetitive motions, aversion to new stimuli, and difficulty with understanding of social situations—was provided in 1925 by Grunya Sukhareva,[49] about two decades before the better known Leo Kanner[50] and Hans Asperger[51] provided their descriptions of autism spectrum disorder.

The year 1980, almost forty years after Kanner and Asperger published their descriptions of autism, marks a turning point in our story. A big medical issue at the time was the connection between Reye's syndrome, a serious condition that causes swelling in the liver and brain, and the use of aspirin to treat the symptoms of viral infections. The concern solidified in 1982 and 1983. This led to a switch from pediatric use of aspirin to acetaminophen, which means that children born on or after 1980 were more likely to be exposed to acetaminophen at an age when they were susceptible to regression into autism. According to the Autism Research Institute, the birth year 1980 was the time stamp that marked the beginning of increased regression into autism.[52]

Another turning point in the scientific understanding of autism occurred in 2005, when "regression" into autism was confirmed by scientists.[53] For decades, parents had been reporting regression into autism in children up to several years of age.[54] Regression involves the loss of previously acquired social, physical, and/or intellectual abilities leading to a diagnosis of autism. Some children, on the other hand, who are diagnosed with autism do not lose any abilities, a condition termed *infantile autism* because autism is apparent from infancy.

I have been asked whether acetaminophen was available before autism existed. It's impossible to know if cases of autism existed prior to 1925, when the first clinical description of autism was published.[55] A much more answerable, and therefore useful, question is whether autism was truly rare in children born prior to 1980, who are less likely to have been treated with acetaminophen.

We have pointed out that the prevalence of autism has risen dramatically since 1980, particularly increasing in the 1990s when direct-to-consumer advertising was introduced.[56] On the other hand, a widely accepted argument has been presented that the real prevalence of autism has not changed extensively over time. This argument is associated with the view that autism is genetic. Indeed, there can be no genetic pandemic. Genes take many generations to change in a population,[57] so if autism were strictly genetic, its prevalence couldn't rise drastically in a mere century. In this view, the apparent increase in the prevalence of autism is largely due to changes in diagnostic criteria, increasing awareness, funding allocations that encourage diagnosis, and other factors such as decreases in social stigma associated with autism. However, this view is relatively easy to dismiss with an objective look at the evidence. What follows is a list of very difficult-to-explain observations if indeed there were no real change in the prevalence of autism since 1980. The list involves both "types" of autism, infantile and regressive. The list is not comprehensive, and some of these observations will be described in more detail later in this book.

- Dramatic correlations between the amount of acetaminophen present at the time of birth and the presence of autism in the resulting offspring were identified in 2020.[58] This observation is supported by studies showing that circumcision, a procedure often associated with acetaminophen use, is associated with infantile autism[59] and with social dysfunction.[60] Further, as described later in this chapter, the period immediately after birth is the time when humans are least able to tolerate acetaminophen.
- Abundant evidence indicates that autism is very rare in parts of the world with chronic shortages of modern medicine. (Details in chapter 4.)
- Investigators working through the 1980s[61] and in the early 2000s[62] saw increases in autism prevalence with either no discernible change in their awareness or methodology or with changes in methodology that could not account for the increases they observed. (Details in chapter 4.)

- Grunya Sukhareva and Leo Kanner were both renowned for their breadth of knowledge and grasp of pediatric psychology, and both described autism essentially the same as it is described today. Yet neither recognized a high prevalence of the disorder in the 1920s–1940s. Furthermore, the scientific community tends to aggressively correct colleagues who make mistakes. Yet Sukhareva and Kanner were not discredited by their contemporaries for describing autism as a novel and rare condition.

- Stephen Schultz found a twenty-fold greater prevalence of regressive autism associated with acetaminophen use in early childhood.[63]

- Although autism appears to be heritable based on sibling and twin studies, the genetics that dictate autism do not exist. This is known as "missing heritability" by scientists,[64] and indicates that the environment present during pregnancy and early childhood plays a critical role in the development of autism.

- Numerous studies in laboratory animals demonstrate that acetaminophen is a potent developmental neurotoxin that affects males more than females. No viable explanation exists for the idea that the drug is safe for developing humans despite being dangerous for developing rats and mice.

Some of the evidence above involves the standard arguments regarding the increased prevalence of autism. The issue with low autism prevalence in areas without modern medicine has been particularly controversial, although some observations from Central America nicely resolve any unknowns there (see chapter 4). However, the observation in the first bullet point above, from a Johns Hopkins study using umbilical cord blood, linking acetaminophen use at birth to autism is equally as compelling.[65] If autism were truly ancient and the prevalence has not changed, why should acetaminophen levels at the time of birth so strongly correlate with autism? It could be argued that somehow, women who are carrying a baby with autism are more likely to be given acetaminophen. In other words, maybe the presence of autism in the fetus somehow did something that caused the physicians to give more acetaminophen. However, acetaminophen does not provide effective control of the pain of labor and delivery and is typically given then only as a matter of protocol. Spinal blocks work extremely well for pain relief during labor and delivery, with acetaminophen providing only a moderate decrease in the need for other, more effective pain relief.[66] In other words, acetaminophen is in the script, so we give the drug even though it only provides some auxiliary pain relief. It's the same as actors following a script, even when everybody in the audience thinks, "*Wow, I would not have done THAT . . . . Why walk into the dark room with the scary music playing and not turn on the light?*" So, in the end, we are left with a quandary. Why should acetaminophen use during labor and delivery be associated with an ancient condition?

We could argue that since many of the cases of autism in the Johns Hopkins study came from different medical centers than many of the controls, it's possible that the medical centers who treated patients with autism had different protocols that included more acetaminophen. In other words, maybe the association found by Johns Hopkins researchers was some sort of weird accident of circumstance? That's possible, but the researchers at Hopkins saw what is called a *dose-dependent response*. This means that high amounts of acetaminophen were much more strongly associated with autism than low amounts, and medium use of acetaminophen had medium risks. That's a clue that something real might be going on. Then we remember that we also have data from circumcision (two independent studies with two independent populations) and pharmaceutical considerations that point to the time immediately after birth as the most critical period. In addition, we have all the other evidence on that list above and much more evidence that will be described later in this book.

In modern science, we tend to pick apart individual studies, which is a wise thing to do. It helps us be critical of our research, and sometimes it can change how we think about the system we are studying. About two decades ago, I conducted a series of studies involving a medical reagent commonly used in surgery.[67] The reagent is called thrombin, and it often comes from cow blood. Yes, surgeons often put molecules from raw (uncooked) cow blood into your body during surgery to help your body recover from surgery, but please try not to think too much about that right now. In my laboratory, we had done an experiment that made us think the thrombin stimulated human immune cells to produce a molecule we really didn't expect those cells to produce. That immune molecule, called *immunoglobulin A* (IgA), is normally found mostly in the gut and in milk, and we had no idea why the thrombin would cause cells to make something that's mostly found in milk.

A helpful scientist pointed out to me that maybe, just maybe, the thrombin was contaminated with that IgA, and the human immune cells didn't produce it at all. Maybe the IgA just came along for the ride, and my colleague suggested that we should run the experiment to find out. There was no apparent reason for IgA to be there, but we were amazed to find that my colleague's wild hunch was absolutely correct. We found that the IgA, normally found in milk, is present in very high amounts in the thrombin preparation. How it got there, we do not know, but we did learn that our immune cells did not produce it. In the end, we learned something that might be important, but the conclusions were not what they first appeared to be. Science can be tricky, and we scientists need to be careful with the interpretation of each experiment.

But, at the same time, if we don't look at the big picture, that picture can be missed. We can argue endlessly about some detail that is difficult to resolve—for example, do the Amish really use less acetaminophen and have less autism?—when in reality the picture is quite clear from a broader perspective. The argument over details, then, is moot when viewed from a larger perspective. In a nutshell, this book is about how scientists missed one of the most important "big pictures" in the drug and medical world of all time.

While we know, based on the deductive reasoning described above, that autism could *not* have existed at present levels prior to 1980, it is also evident that changing awareness and improved diagnosis, along with increased financial incentives for an autism diagnosis, have certainly increased the diagnosis of autism. If it were possible to bring Grunya Sukhareva and Leo Kanner through time into a clinic today, we might speculate that they would miss perhaps 50 percent, give or take 20 percent, of the cases of autism that we recognize today. We can further speculate that they would probably identify classic cases in the middle of the spectrum, which they accurately described in their publications, but they might fail to detect those very severely and very mildly affected. Drs. Sukhareva and Kanner might also fail to detect unusual cases that we might today classify as autism. What this means is that their estimates of autism in 1920 or 1940 might be lower by as much as a factor of two or even three compared to what we would consider accurate today.

With that in mind, using a rough estimate of one in 2,500 for measures of autism in the 1940s, we could speculate that the prevalence of autism was as high as one in about 800 or so in the 1940s if we use modern diagnostic criteria and methods. Today's prevalence, one in 31, is still more than 25 times greater. If the prevalence of autism has remained steady, Drs. Sukhareva and Kanner would have missed the vast majority of all cases that would be identified using today's methods. This situation would be difficult to explain given their published descriptions of autism, which closely resemble the diagnostic criteria we have today. Moreover, the other evidence described above defies explanation if they did indeed historically miss the vast majority of cases. By way of reminder, the conclusions we have reached do not depend on this one hypothetical thought experiment that can never be conducted in reality. It depends on a larger body of evidence. No one bit of evidence is conclusive. It's the weight of total evidence that we care about.

Looking at the bigger picture, it seems unlikely that, if the classic symptoms of autism were as common as they are today, two renowned clinicians would believe they had found a new and rare condition when they precisely described the symptoms we recognize today as characteristic of autism. In this rather fantastic scenario, the rest of the clinicians at that time must have also been completely oblivious.

Some have assumed that if we can detect twice as much autism today as they could in 1920 or 1940, then only half of today's autism prevalence is due to a real increase. The math doesn't work that way. Rather than dividing today's prevalence in half, we are supposed to double the prevalence they found in 1920 or 1940. If we do that, we see that most of the increase in the prevalence of autism is real, and then everything makes sense—all of the available information points in the same direction.

After children's exposure to acetaminophen began increasing in 1980, our story took another dramatic turn twenty-eight years later, when Stephen Schultz published his study finding that exposure of young children to acetaminophen was associated with a twenty-fold increase in regressive autism. The study was largely ignored,[68] but a careful analysis reveals no substantial flaws.[69] The study was a *case-controlled study*, involving about eighty children with autism and a similar number of neurotypical children. It seems likely that the study was biased toward individuals who believed their child's autism was induced by a vaccine,[70] a situation that would not affect the conclusions we draw from the study. The fact that many parents associate their child's autism with vaccination does not affect our conclusions, since vaccination is often associated with acetaminophen exposure. This issue will be covered in some detail in chapter 5.

The reason the 2008 Schultz study is pivotal is that it was the first to propose the idea that acetaminophen could be critical in the development of autism. Until 2008, the idea simply did not exist in the public consciousness. If the study had been taken seriously, a broader look at the wider evidence would have revealed a wide range of extremely concerning evidence. That evidence, available in 2008, includes but is not limited to the following.

- The increase in the prevalence of autism coinciding with the switch from aspirin to acetaminophen was evident by 2008.[71] In addition, advertising of acetaminophen was intense during the 1990s and 2000s, with between 115 and 250 million dollars (206–488 million dollars/year with inflation adjustment for 2025) spent per year.[72] By comparison, a major blockbuster movie may have an advertising budget of 150 million dollars in 2025.

- The fact that acetaminophen is converted into a toxic substance by the human body under certain circumstances was known in the 1970s,[73] and the fact that many children with autism could not safely process the drug was known by the 1990s.[74] See Figure 1.1.

- The observations of parents connecting the measles, mumps, and rubella vaccine with autism had already been described by 2008.[75] Since acetaminophen is used frequently with vaccination, the role of acetaminophen in the induction of autism provided a viable explanation for the parent's observations regarding vaccines. Ignoring parents' observations had, historically, proven to be disastrous for the medical establishment.[76]

- It was already known that newborns were deficient in a metabolic pathway necessary to process acetaminophen safely. Veterinarians also knew that domestic cats could not tolerate acetaminophen[77] because they were deficient in the same pathway.[78]

- It was known in the 1980s that even lethal doses of acetaminophen did not damage the livers of neonatal laboratory animals,[79] yet an analysis of the literature at that time would have revealed that the *belief that acetaminophen was safe was based strictly on liver function.*[80] Despite the fact that acetaminophen targets the brain, profoundly affecting several molecular systems vital to brain function, the drug was never tested for safety during brain development.[81] In fact, acetaminophen was not directly tested for safety during neurodevelopment until 2014, six years after the Schultz study was published. It spectacularly failed that test when acetaminophen's effects on brain development in newborn laboratory mice were finally studied.[82]

The Schultz study, combined with knowledge available in 2008, should have been enough to change clinical practice. Even if, at that time, the evidence was not yet incredibly *convincing*, it was more than *concerning*. Figure 2.1 depicts what is currently known about the metabolism of acetaminophen in children, and especially children with autism. Most of that information was known in 2008.

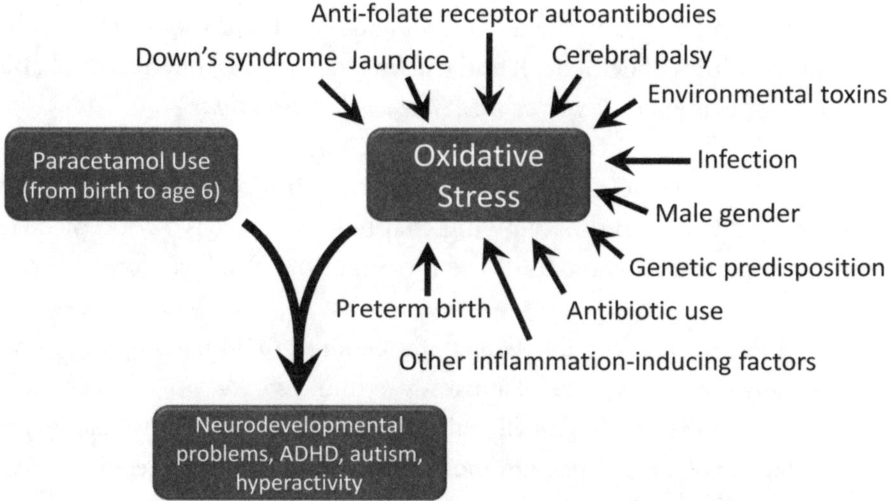

Figure 2.1: The metabolism of acetaminophen is impaired under some circumstances. The diagram was modified from a similar diagram published in *Clinical and Experimental Pediatrics,*[83] the flagship journal for the Korean Pediatric Society. The chemical name for acetaminophen is *N-acetyl-p-aminophenol*, which is abbreviated *APAP*. The toxic molecule that the body makes from acetaminophen is *N-acetyl-p-quinoneimine*, which is generally abbreviated *NAPQI*.

At the time of this writing, the evidence pointing toward the importance of acetaminophen in the development of autism has mounted far beyond the evidence present in 2008. In our most recent work, we now count about thirty distinct lines of evidence—although exactly how it is divided into the thirty lines is a bit arbitrary. The evidence could easily be grouped into twenty-five lines of evidence if we combine things more aggressively, but we could also get a few more lines of evidence out of it, especially if we divide up the biochemical and pharmacological evidence more finely.

At this point, I will refer the interested reader to Appendix B, which contains a summary of scientific evidence telling us that exposure of susceptible babies and children to acetaminophen triggers many if not most, and probably the vast majority of, autism cases. It's not important to understand the details for now, but I would like the reader to have an overall impression of the strength of total evidence. For the sake of presenting the evidence, I have divided it into four categories, separated into four tables in Appendix B. The first category is *pharmacologic evidence*, related to how acetaminophen interacts with the human body (Appendix B, Table 1; 6 lines of evidence). The second category *includes associations between acetaminophen use and autism through time, place, or human activity* (Appendix B, Table 2; 12 lines of evidence). The third category of evidence is derived from *laboratory animal studies* (Appendix B, Table 3; 4 lines of evidence), and the final category (Appendix B, Table 4; 8 lines of evidence) involves *miscellaneous observations*.

Appendix B very briefly summarizes a considerable volume of evidence, and we are going to take a deep dive into that evidence in an understandable way in the following chapters. But here is a good place to review our conclusions stated in the introduction. We have five of those.

1. Exposure of susceptible babies and children to acetaminophen causes many if not most cases of autism spectrum disorder.

2. The best explanation for all available evidence is that the vast majority of all cases of autism, perhaps more than 90 percent, is triggered by exposure of susceptible babies and children to acetaminophen.

3. Susceptibility is caused by a complex array of genetic and environmental factors that lead to *oxidative stress*, a condition in which the body cannot keep up with the chemical burden it faces.

4. The time of greatest risk for brain development is during and immediately after birth. Risk diminishes within weeks and dissipates almost completely by the time the child reaches six years of age. Some risk during pregnancy seems likely, but with less risk and with less certainty of risk than at birth.

5. Three rudimentary scientific errors—combined with subversion of the scientific process—have blinded most clinicians and scientists to the connection between acetaminophen and autism.

These conclusions are, of course, based on the total body of evidence summarized in Tables 1 through 4 in Appendix B rather than any one particular line of evidence. Some of that evidence is very, very extensive, especially the part about evaluating alternative explanations. The summaries in the tables do not display all of that complexity. Importantly, the conclusion that "acetaminophen causes autism" is not technically correct, any more than is the statement "codeine causes respiratory failure and death." In both cases, with acetaminophen and with codeine, *susceptibility to injury is required*. In the case of codeine, a single enzyme is the problem. If that enzyme is overactive, the child's body converts codeine into morphine too quickly, and the child can die. With acetaminophen, the picture is much different. You've seen in Figure 2.1 just some of the complexity of the way our body deals with acetaminophen. In reality, many enzymes, vitamins, and other molecules are involved in those pathways. The result is that sensitivity to acetaminophen depends on many factors, including the child's phase of development, hundreds of genetic factors, and many dozens of environmental factors. The combination of genetics and environment that we call *epigenetics* is also important. A diagram explaining the concept is shown in Figure 2.2.

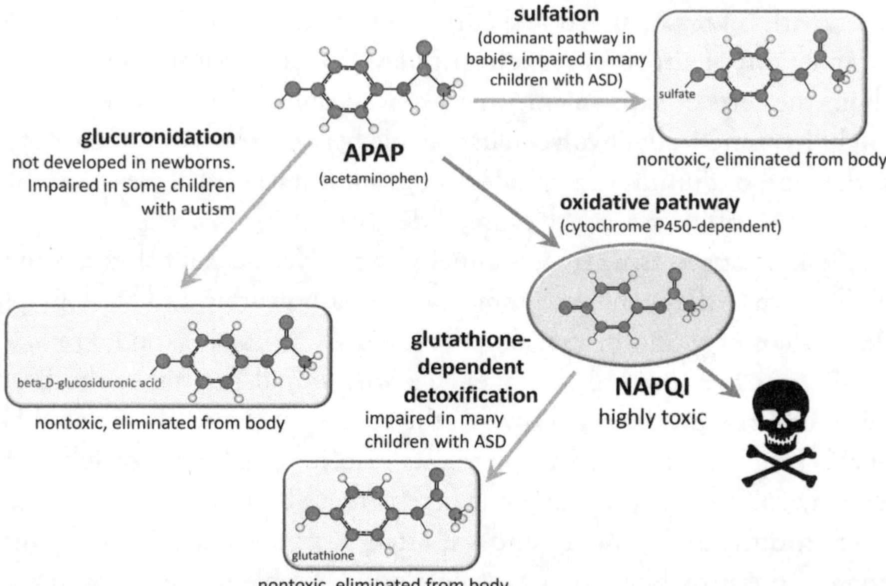

Figure 2.2: The interaction between acetaminophen (paracetamol) and oxidative stress, leading to neurodevelopmental injuries. This figure was published in the journal *Minerva Pediatrics*,[84] the oldest pediatric journal in Italy. Since the journal is European, the drug name paracetamol was used rather than acetaminophen, the name which is used in the USA.

I've been asked, *How confident are you in those conclusions?* The answer we have published is that we have no reasonable doubts about the conclusions, although that probably only applies to the first three conclusions, not the fourth. If I had to put a number on it, I would say 99.99 percent sure, but other members of my team would probably give different numbers. If we stopped using acetaminophen from conception to age six—and especially during labor and delivery—my best guess on the resulting reduction in the prevalence of autism is 94.5 percent, but other members of my team have different numbers in mind. Finally, the fourth conclusion has a lot of wiggle room, and additional data is needed if we want more clarity, particularly about the risks of acetaminophen use during pregnancy. The fifth conclusion is as much an observation as it is a conclusion. I'll present evidence for that conclusion throughout the pages of this book.

The bottom line is that we don't have absolute proof that acetaminophen plays a role in the induction of autism. In a similar way, nobody ever proved that parachutes save lives, because that might be difficult, not to mention ethically dubious. But since we know that a parachute slows vertical descent, and we know that high-speed impacts are lethal, we can infer, with no reasonable doubt, that parachutes do save lives for people jumping out of airplanes at high altitudes. We don't want to even think about the sorts of experiments that would give us absolute proof, which might hypothetically involve many people being thrown out of airplanes with placebo (dummy) parachutes under anesthesia so that we could rule out the fear of impact as the cause of death.

Looking at this issue from another perspective, we can't shrink somebody down so they can take a camera into a carbon atom and photograph the nucleus of the atom to make sure it has six protons. Who knows . . . maybe our hypothetical shrunken scientist would get hit in the head by an electron on the way down to the atomic nucleus, which could be fatal? (I ask that the nuclear physicists reading this book not take this analogy too seriously.) The point here is that absolute proof in science is often difficult to obtain, and we often accept scientific "facts" with good justification but without absolute proof. Problems happen when we assume something to be true without actually looking at the underlying evidence. That's what happened when acetaminophen was assumed to be safe for use during brain development. Incorrect assumptions were made, and my research team at Duke University proved that the assumption of safety was based on faulty reasoning.[85] In this case, we did obtain absolute proof that acetaminophen was never proven to be safe, and we now include that in our lines of evidence (line of evidence #23 in Appendix B, Table 4). So, while we have absolute proof for some lines of evidence, and we are confident without reasonable doubt about the main conclusions, we do not have absolute proof of the main conclusions.

In the next chapters, I'll dig into the evidence in more detail. This is where we will see the scientific mistakes that led to the current confusion about acetaminophen's connection with autism.

Chapter 3

THE SINGLE MOST BLINDING SCIENTIFIC ERROR: THOSE ARE NOT CONFOUNDING FACTORS

After President Donald Trump and Secretary of Health and Human Services RFK Jr. made an announcement that acetaminophen causes autism, a plethora of papers in reputable scientific and medical journals claimed that no reliable evidence pointed toward any danger of acetaminophen use during pregnancy.[86] Some of the most prestigious medical and scientific journals in the world, including *Nature*,[87] the *Journal of the American Medical Association* (*JAMA*),[88] and *The BMJ*,[89] were among the offenders. Ironically, Baylor Medical Center sent me a letter explaining that acetaminophen doesn't cause autism, as "some people" were claiming. I was on their mailing list because my sister had worked there.

Here I should hasten to point out that sometimes scientists dismiss concerns about acetaminophen without actually saying that "acetaminophen is safe" or that "it does not cause autism." Not technically, anyway. In technical terms, they sometimes only say there is no real association between acetaminophen use during pregnancy and autism, ADHD, or other neurodevelopmental problems. But scientists and doctors know that, without association, there can be no causation. If Joe Banana the criminal was in New Mexico when a burglary happened in France, then Mr. Banana was *not* the burglar who physically took the jewels in France. By the same token, when a prestigious paper was published in 2024 with the absolutely wrong conclusion that "Acetaminophen use during pregnancy was not associated with children's risk of autism,"[90] an administrator at the US National Institutes of Health followed up with a press release entitled "Study Reveals No Causal Link Between Neurodevelopmental Disorders and Acetaminophen Exposure Before Birth."[91] All would agree, if there is no association, there can be no causation. But, as we will see in this chapter, abundant associations exist and are simply ignored for a reason that is verifiably wrong.

Some of the important evidence we have are associations between acetaminophen use and autism. We see these associations during three distinct time periods: pregnancy, at the time of birth, and during early childhood.

The easy part of this chapter is understanding the rudimentary error that has caused most of the greatest minds in the world, almost without exception, to overlook the association between acetaminophen and autism. It's not much more difficult than understanding how doctors' refusal to wash their hands, wear clean clothing, and provide clean linens for patients before labor and delivery was a big killer of women in parts of Europe during the mid-1800s.[92] Despite solid evidence produced by the legendary clinician Ignaz Semmelweis, many of the great medical minds of the mid-1800s refused to wash their hands before attending deliveries, and refused to ensure that clean clothing and linens were used. The scientific error affecting our understanding of the connection between acetaminophen and autism is just as easily understood and will be explained in this chapter. The harder issue to comprehend is, how is it possible that this problem exists? The whole situation sounds unbelievable. How could a fiasco of this magnitude have happened?

In attempting to come to grips with the presence of an incomprehensively damaging error in our midst, I can only offer that history is full of examples of the most prominent members of the medical community being misled. Doctors performed surgery for years without washing their hands, causing thousands of deaths in the 1800s, prescribed thalidomide, which caused thousands more deaths as well as birth defects in the 1900s, and were crucial participants in the opioid epidemic in the 2000s. By analogy, doctors are much like astronauts. Astronauts are some of the most highly trained people on the planet, but they strictly depend on the engineers who design their technology. If the technology is flawed, their space shuttle can, tragically, become a death trap. By the same token, if the science underlying a drug is flawed, that drug can cause incredible amounts of damage, and physician training will not help. In this chapter you'll see that experts writing in prominent medical journals have made rudimentary and catastrophic scientific errors when it comes to acetaminophen.

Before we look at the scientific error currently blinding us to the connection between acetaminophen and autism, one more detail should be considered. Some fully trained scientists in the field of *epidemiology* have been involved in the rudimentary error. The error itself occurs at the intersection between *pharmacology* and *statistics*, two branches of science distinct from the field of epidemiology. Although epidemiologists, who deal with the prevalence, distribution, and prevention of disease at the population level, have significant amounts of training in statistics, they are not statisticians. And they have essentially no training in pharmacology. It is within the rudimentary principles of statistics and pharmacology that the catastrophic error lies. In our astronaut analogy, the engineer who devised the guidance computer will be unaware of problems regarding the stability of the seals on the fuel tanks in cold weather. If we want to know about the safety of the seals on the fuel tanks, we must ask the engineer who knows everything about the material that makes up those seals. That's how we lost the *Challenger* Space Shuttle and its entire crew on January 28, 1986. Nobody listened to the engineer who was trying to point out that the fuel tank's seals wouldn't work in cold weather—a fatal flaw. With that flaw, no amount of astronaut training could have saved the *Challenger*, and no amount of brilliant engineering in the guidance system was going to make a difference in the outcome. Knowledge from the right engineer, or the right scientist, is sometimes critical for avoiding disaster.

To begin to understand the critical scientific error blinding some of the world's most brilliant minds to the connection between acetaminophen and autism, it is first important to understand the difference between *correlation* and *causation*. Correlation means that two things go together in some way or another. The presence of coffee cups, for example, is correlated with the presence of coffee. This means that we see more coffee cups in places where coffee is present than in places where coffee is absent. We can, of course, find coffee cups in places without coffee, but if we search the entire world over, we will find that the presence of coffee cups is, on average, correlated with the presence of coffee. But coffee cups do not "cause" coffee to exist. Correlation does not equal causation. In this case, both the presence of coffee cups and the presence of coffee have an underlying cause. Humans like to use coffee cups to drink coffee. Our species likes to keep the two items together for convenience, leading to the correlation.

Telling a professional scientist that correlation does not equal causation is equivalent to telling a professional American football player how to distinguish between a cheerleader and a player eligible to catch the ball during a game. We don't need to be concerned that a professional football player will intentionally throw the ball to a cheerleader and expect her to catch it and run it down the field, avoiding tacklers, in an attempt to score points. Cheerleaders and football players are easily distinguished. For example, cheerleaders don't wear helmets, and all the players do. Since the cheerleaders don't wear helmets, it wouldn't be safe for them to play in the game, removing any temptation to throw the football to a cheerleader in the first place. The point of this outlandish example is that we don't need to worry that a professional scientist is going to confuse correlation with causation. I'll point out here that, for scientists, "association" is probably a better term to use than correlation in this general context. You'll see me use the word "association" throughout this book, and for practical intents and purposes, association means the same thing here as correlation. So, associations do not equal causation.

I have not attempted to count the number of times somebody has pointed out to me that associations don't equal causation, believing that they have come up with a brilliant argument that somehow overcomes over a decade of my team's work. I recall one such incident from about a decade ago in the ivory tower at Duke. I pointed out to a freshman undergraduate student that circumcision, a procedure associated with acetaminophen use, usually at the time of greatest risk from acetaminophen-induced injury, was associated with autism. (This evidence will be covered in detail in chapter 5.) He immediately gave me a stern warning that correlation does not equal causation. I asked him if he had an explanation other than causation for the association between acetaminophen use and autism in this case, which of course he did not. Apparently, he had not considered that associations can happen *because* of causation. Neither could he explain numerous other lines of evidence I provided to him which pointed toward acetaminophen as a trigger for autism. Remember Stephen Schultz from chapter 2? He saw his son regress into autism, and his data revealed a very large association between regression into autism and the use of acetaminophen in early childhood. At that point, he basically had no other suspects upon which to blame the regression, unless regression was "spontaneous." He was literally out of ideas regarding what else could be tested, although he could not rule out some factor that he was ignorant of. The idea here is that associations can sometimes be due to causation, but we need to look at the bigger picture before we can draw a conclusion.

When looking at the bigger picture, we ask questions such as, *how many associations are there*, *do any potential confounding factors exist* (explained later in this chapter), *what evidence exists other than associations*, and *what other suspects do we have*? If you enjoy reading or watching murder mysteries, you'll probably recognize something familiar. Scientists think much like detectives when looking at evidence. Scientists often look at the body of evidence and use deductive reasoning. For example, the concept that the simplest solution that explains all of the observations is likely the correct solution, called Occam's razor, is an important part of this discussion and is an integral part of scientific reasoning.[93] Further, Occam's razor can be extended to include the idea that the greater the difference in simplicity between two solutions that explain all the observations, the greater the likelihood that the simpler solution is correct. The conclusions reached in this book are not based only on associations between acetaminophen use and autism, or on any other single bit of evidence, but on the total weight of all evidence, evaluated using scientific principles such as Occam's razor.

The error described in this chapter leads scientists to conclude that the associations between acetaminophen and autism are not valid. This error is incredibly blinding because another scientific principle is that, without associations, there can be no causation. Joe Banana, our burglar, helped us illustrate this scientific principle earlier. Therefore, scientists conclude that, since acetaminophen use is not associated with autism, acetaminophen cannot be involved in the induction of autism. A scientific principle, in this case, leads to the wrong conclusion because of an error with the information leading to that conclusion. But science has a way of uncovering such errors. When a conclusion based on one line of evidence contradicts multiple other lines of evidence, we know that examination of our underlying assumptions is necessary. When we perform that examination, the error is obvious.

In our earlier example, the space shuttle *Challenger* was launched in cold weather using fuel tank seals made of material that becomes inflexible in cold weather. This situation was associated with tragedy. In-depth investigation determined that this association was indeed causal. The failure of the seals due to cold temperatures caused the fuel leak that led to the explosion. A common question scientists ask is, *How do we know if two associated factors are connected by a causal relationship*? To evaluate causal relationships, we need a detailed understanding of the system we are studying, and things called *confounding factors* are often important.

The best way to explain the concept of confounding factors is with some examples. If we were in a war zone, we might see an association between the number of bandages present on a body and death. The more bandages a soldier's body has, the more likely that soldier is to be deceased. With that observation, we could (very) naively conclude that bandages cause death. But here we have an obvious confounding factor. More bandages equals more *wounds*, and more wounds equals a greater risk of death. In other words, the bandages did not cause death. Association, in this case, did not equal causation. Wounds obtained during combat were a confounding factor that caused the association between bandages and death. A less trivial example, that I can provide peer-reviewed scientific reference for, is the association between coffee drinking and lung cancer. The association is actually real:[94] the more coffee the average person drinks, the more likely they will eventually be diagnosed with lung cancer. We might suspect from this that coffee causes lung cancer, but as you may know, coffee contains lots of antioxidants, and is generally considered to be fairly healthy.[95] The confounding factor here is cigarette smoking. People who drink lots of coffee *also* tend to smoke.[96] So, smoking, like wounds on a battlefield in our previous example, is the confounding factor responsible for the association between two things. Coffee does not cause lung cancer any more than bandages cause death. Association did not equal causation, and the association was due to the presence of a confounding factor. Figure 3.1 shows how coffee drinking, smoking, and lung cancer are connected. This diagram will be important later when we consider acetaminophen.

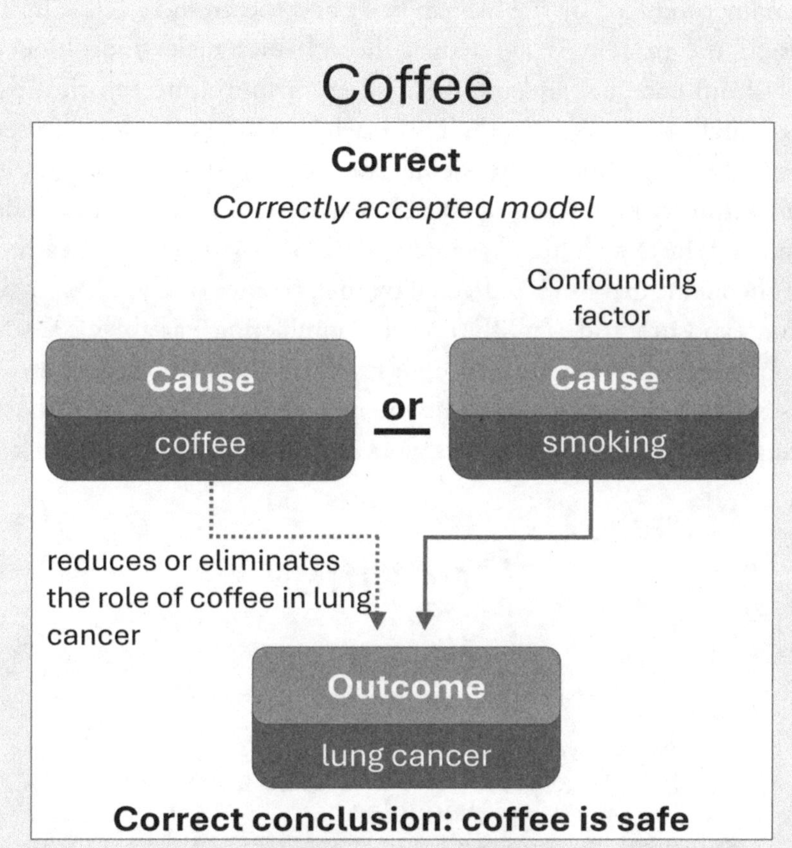

Figure 3.1: Confounding factor. Associations between coffee drinking and lung cancer might trick some people into thinking that coffee drinking causes lung cancer. But scientists correctly determined that a confounding factor was responsible for associations between coffee and lung cancer. Smoking, not coffee, causes lung cancer.

There are three important scientific concepts required to fully understand the error that has tragically confused some of the world's most renowned scientists with regard to acetaminophen with autism. The concept of associations and causation is the first, and the concept of confounding factors is the second. The third and final concept is *interacting variables*.

Interacting variables are variables that interact to create an outcome. One of the best-known examples is the interaction between a drug called codeine and a protein that converts codeine into morphine. Codeine is widely used for pain control, and until only a few years ago, was widely used in babies and children. Codeine itself doesn't do anything useful, but when the body converts codeine into morphine, pain is readily controlled. The protein that converts codeine into morphine, called CyP2D6, is naturally produced by the human body and is extremely active in some babies. If the protein is *too* active, those babies make morphine very quickly from codeine, and the rapid surge of morphine can shut down their respiration, causing death. This is why codeine is no longer used in babies and young children. Codeine can cause death—but only in individuals with very high levels of CyP2D6 activity. In other words, codeine exposure interacts with high levels of CyP2D6 activity to cause death. The effect of codeine exposure is affected by the presence of CyP2D6, making codeine exposure and CyP2D6 levels "interacting variables." Codeine at therapeutic levels is safe, and high levels of CyP2D6 activity are safe, unless the two come together. The interaction can be deadly. A diagram showing how interacting variables work is shown in Figure 3.2.

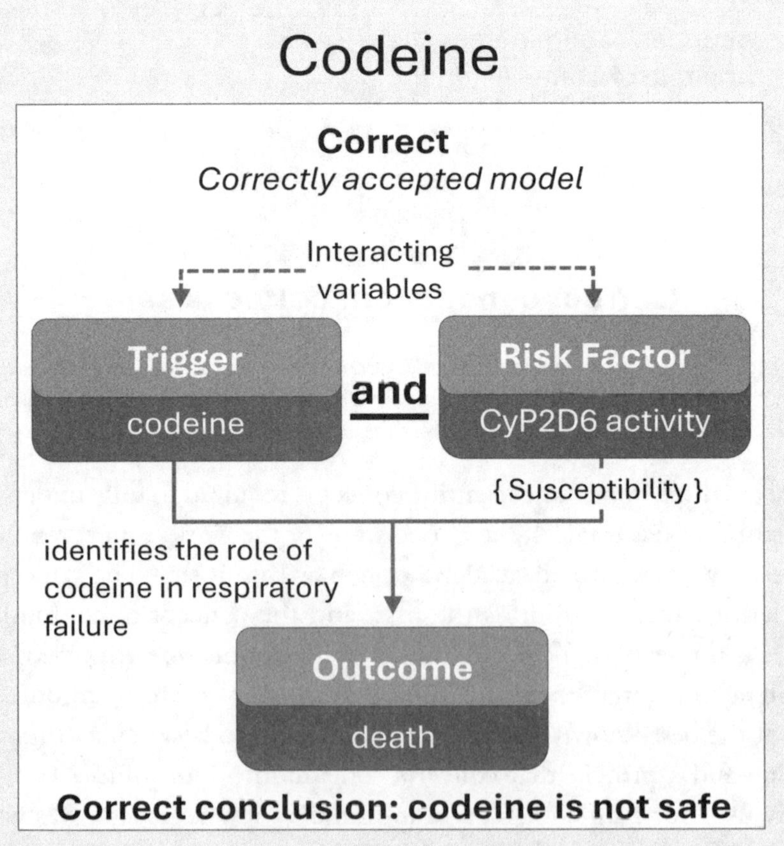

Figure 3.2: Interacting variables. Two factors, codeine and an enzyme called CyP2D6, interact to cause respiratory failure and death in some babies and young children with high levels of CyP2D6 activity.

So, what is the big problem preventing most great scientists of our day from seeing a connection between acetaminophen and autism? The answer is simple: when it comes to acetaminophen, *they have assumed that interacting variables are confounding factors*. Unfortunately, and even tragically, scientists have decided that acetaminophen is like coffee, not codeine. They have decided that, like coffee, acetaminophen is safe, and something else (like cigarette smoking) must be the problem, in this case causing autism. But other factors interact with acetaminophen to make it more or less dangerous.[97] In a nutshell, we *know* that acetaminophen is more like codeine than coffee. Figure 3.3 illustrates the difference between current assumptions about acetaminophen and reality.

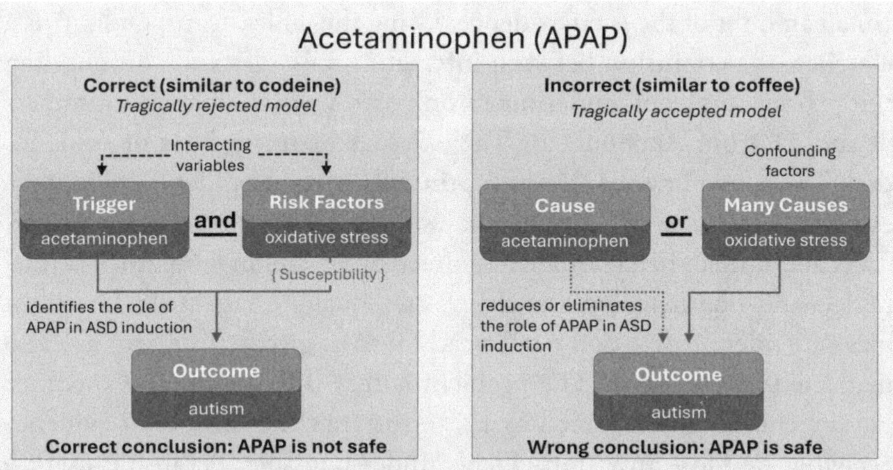

Figure 3.3: Confusing interacting variables with confounding factors. In the case of acetaminophen (known by scientists as APAP), the tragic assumption was made that acetaminophen is like coffee; thus, any association with autism (ASD) is due to confounding factors. Abundant and robust evidence published in the peer-reviewed literature and described in this book tells us this assumption is wrong.

We understand that coffee is safe and that smoking is the confounding factor that, by itself, leads to lung cancer. Tragically, many great scientists believe that acetaminophen is safe because they believe that oxidative stress is a confounding factor that, by itself, somehow leads to autism or is associated with unknown factors that lead to autism. Many things can cause oxidative stress, and all these things would also be considered confounding factors, rather than the interacting variables they really are. This would be the same as if we decided that codeine does not cause death, but rather a high level of CyP2D6 activity, by itself, causes death.

In that case, we could continue to give codeine to babies, and we would continue to see some of them dying.

Fortunately, scientists figured out that codeine was interacting with high levels of CyP2D6 activity. But with acetaminophen, scientists have missed the big picture and failed to recognize that it interacts with oxidative stress to cause autism. In a nutshell, scientists in 2013 decided that factors leading to oxidative stress were confounding the connection between acetaminophen and autism.[98] Those scientists were apparently unaware of abundant evidence demonstrating a causal relationship between acetaminophen, oxidative stress, and autism.

Before I move on to explain exactly how the error happened and the consequences of that error, I want to point out that the error only affects a small amount of the total evidence. Using the tables in Appendix B as a reference, the confusion between interacting variables and confounding factors has completely undermined only two lines of evidence, numbers 10 and 11 from Appendix B, Table 2, out of thirty lines of evidence total. Those two lines of evidence primarily involve epidemiologic studies of data from health-care databases. The other associations in time, place, and human behavior are not affected. Studies in laboratory animal models and pharmacologic evidence are equally unaffected. The eight lines of evidence that don't fall neatly into a specific category are also unaffected by the error. The problem is that the two lines of evidence that are entirely undermined by this error are the two lines of evidence receiving the most attention. That's understandable, because causation can't happen without association. If the data really are saying that there is no association, then there is no point in looking at the issue further—a tragic but understandable error.

Remember the diagram of acetaminophen processing in Figure 2.1? If the acetaminophen isn't safely processed, it is metabolized to a toxic compound called *NAPQI*, which can cause damage and cell death. As the diagram indicates, a molecule called *glutathione* is necessary to safely process acetaminophen in our body. Without it, levels of NAPQI increase. We also saw in chapter 2 that some individuals, especially children with autism, are missing or deficient in other metabolic pathways that help process acetaminophen safely. Babies who have low glutathione or are missing other pathways that help metabolize acetaminophen *will be susceptible to acetaminophen-mediated injury*. This is not speculative, or simply a hypothesis—or in common terms, a theory. This is a *known* fact of the branch of science we call pharmacology: acetaminophen gets more dangerous if we have low glutathione or other problems related to acetaminophen metabolism. Unfortunately, our body's system for processing acetaminophen is not as simple as the single enzyme that processes codeine to make morphine.

Glutathione is called the "master antioxidant" and is used by everything from humans to pine trees to mushrooms to deal with numerous stresses that our bodies encounter. With the exception of some bacteria, glutathione is fundamental to life on this planet. Many factors can cause chronically low glutathione levels or drain our glutathione levels, putting us into a condition known as "oxidative stress." With that in mind, you may not be surprised that many, many genetic and environmental factors affect our glutathione levels, putting us at more or less risk for toxicity from acetaminophen.[99] Everything from having an infection[100] to being emotionally depressed[101] can negatively impact glutathione levels.

Unfortunately, the authors of the 2013 paper that first assumed that interacting variables were confounding factors[102] underestimated the association between acetaminophen use and neurodevelopmental problems because they did not take into account that oxidative stress interacts with acetaminophen to cause toxicity. It seems very likely to me that these scientists were either (a) unaware of how the body processes acetaminophen, (b) did not realize that the factors they considered to be potentially confounding would affect oxidative stress, and/or (c) were unaware that the processing of acetaminophen profoundly affects how they should analyze their data. The first two lapses involve pharmacology, and the third is a lapse in statistical knowledge. It is unthinkable that they made the mistake on purpose. Rather, I believe they simply made a mistake in assuming that variables known to interact with acetaminophen metabolism should be treated as confounding factors that do not interact with acetaminophen. The resulting math they and many others after them performed is essentially answering the wrong question. The one they answered is, is acetaminophen dangerous *in the absence of oxidative stress?*

Acetaminophen is not, at least as far as anyone knows, dangerous in the absence of oxidative stress. Acetaminophen certainly doesn't cause autism in the absence of oxidative stress. If it did, almost all children today would have autism, and it would be difficult to explain why so many factors related to oxidative stress are associated with autism.[103] Further, we expect that the toxic effects of acetaminophen will emerge *only* in the presence of oxidative stress based on the known pharmacology of the system, shown in chapter 2. Treating factors related to oxidative stress as confounding factors is akin to asking whether it's safe for children to play with matches if they are wearing fireproof clothing from head to toe and there are no flammable materials in the environment. In those *specific* circumstances, children playing with matches is "safe." But in general, we know that children playing with matches is *not* safe.

While the main problem was treating interacting variables as confounding factors, a couple of other issues contributed to the error. First, acetaminophen use is very common—it's used by many people who are *not* susceptible to injury as well as most people who *are*. If acetaminophen use was rare, the connection between acetaminophen and autism would probably be obvious, even if we were confused about interacting variables and confounding factors. It's important to keep in mind here that *all* acetaminophen use during the period of susceptibility matters. If a study focuses only on pregnancy, for example, it will, by design, miss exposure to acetaminophen during labor and delivery and during early childhood. Evidence that tells us exposure during those times is very important, as will be discussed in chapter 5.

Numerous discussions in the scientific literature and on social media have grappled with whether a particular study accurately assessed all the acetaminophen exposures during pregnancy. Did their survey find all exposures, or did they only measure a fraction of them? But if the study team didn't check exposure during labor and delivery and during early childhood, they missed a lot of important exposures. *No study has ever looked at acetaminophen exposure through the entire period of susceptibility.* This is one reason why we can't rely on a single study to get answers. The only way that we can draw our conclusions is by piecing together available evidence from different sources to form a coherent picture.

Another way to look at the situation is to consider what would happen if everybody, from the smallest child to the oldest adult, smoked cigarettes. If that happened, lung cancer would be very common, but only those who are genetically susceptible would get it. Thus, lung cancer would *appear* to be genetic rather than induced by smoking as we know it to be. The more common something gets, the harder it can be to identify as a problem.

One other problem that contributes to the confusion is that sicker individuals tend to have more oxidative stress, which makes them susceptible to acetaminophen-induced injury. At the same time, because they are sicker, they are more likely to use acetaminophen. In other words, babies and small children more likely to be *injured* by acetaminophen are also more likely to be *given* acetaminophen.

We can sum this catastrophic situation up in a few bullet points.

- Acetaminophen use is very common, and scientists aren't measuring all the potentially damaging acetaminophen use in any study because they are focused on limited time frames such as pregnancy. The situation is made worse because some studies depend on medical surveys that ask people which "drugs" they use, and many people don't think of common over-the-counter medications as "drugs." This means that some studies use inaccurate data. This issue will be examined in more detail in chapter 5.
- Having oxidative stress makes individuals susceptible to acetaminophen-induced injury, but oxidative stress is more often found in people who are relatively sicker, and those people are likely to use more acetaminophen.
- Interacting variables are assumed to be confounding factors in scientific studies. With common heavy use of acetaminophen and with sicker people getting more, standard statistical analyses find that the *reason* for giving the acetaminophen, or other factors related to oxidative stress, not the acetaminophen itself, is causing autism.[104]

A well-known paper was published by a group at Drexel University in 2024 in the *Journal of the American Medical Association* (*JAMA*).[105] This "2024 *JAMA* paper" evaluated autism and acetaminophen use during pregnancy, included more than two million Swedish children, was widely publicized, and was promoted by a director at the NIH using the NIH's official press relations program.[106] The paper confused confounding variables with interacting factors, erroneously concluding that no legitimate association between acetaminophen and autism exists. Many other studies, starting in 2013, made the same scientific error prior to

2024, but the Drexel group had more data and more detail in their data than anybody before them. Profoundly affecting the field to this day, that study, quite simply, executed the error of confusing interacting variables with confounding factors *better* than any other previous study. In other words, the team at Drexel used the wrong technique more effectively than anybody else had ever used it, achieving the "most wrong" answer to date.

Here, in Figure 3.4, you can see some of the actual results from the 2024 *JAMA* paper published by scientists at Drexel. Notice that the amount of uncertainty, indicated by the bars above and below each point of data, is very small. Drexel scientists could be very confident in their answer because they had so much data. More data equals more certainty. But since the wrong approach was used, more data equals more certainty—about a wrong answer.

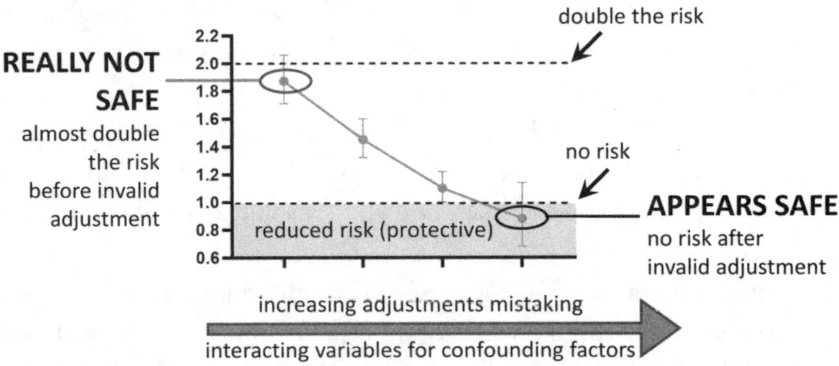

Figure 3.4: The result of treating interacting variables as confounding factors in the 2024 *JAMA* paper published by Drexel University.[107] In this case, data are shown for heavy use of acetaminophen during pregnancy, which typically occurs with treatment of chronic pain, a situation that involves oxidative stress. This graph is reproduced from a video explaining the problem with assuming that interacting variables are confounding factors. That video accompanies the peer-reviewed manuscript[108] our team published explaining the problem with the 2024 *JAMA* paper, which can be found on the publisher's website.

What is obvious in the data from Drexel (Figure 3.4) is that heavy use of acetaminophen during pregnancy is associated with an almost twofold increased risk of autism. This is shown on the far left of the graph. This means that, if a woman uses acetaminophen extensively during pregnancy, then she is almost but not quite twice as likely to have a child who will eventually be diagnosed with autism. But it's also apparent that the association can be "canceled out" if interacting variables, factors known to be associated with oxidative stress, are essentially removed from the analysis by treating them as confounding factors. This can be seen in the second two data points, where increasing numbers of interacting variables are removed. By the third data point, which shows no significant risk, *more than thirty interacting variables* have been treated as confounding factors. In the final data point, on the far right of the graph, the result of removing interacting variables and doing what the authors describe as a "sibling control analysis" is shown. Please keep in mind that the final data point is a combination of incorrectly assuming that interacting variables are confounding factors AND the sibling control analysis. The sibling control analysis by itself didn't affect the analysis to a large degree. The results in the graph were already not statistically significant before the sibling analysis was conducted.

The sibling control analysis performed by the Drexel scientists received great attention in the field of autism research. A sibling control analysis is often used to separate the contribution of genetics and environment. Such an analysis, if done properly, can be very telling. The general idea is to look at siblings who had different amounts of acetaminophen exposure during pregnancy. If the diagnosis of autism depended more on family relationships than on the amount of acetaminophen exposure, then presumably something about being a sibling, NOT acetaminophen, was responsible for autism or lack thereof. This might be a brilliant way to look at the data, in theory. But this is not what the Drexel team actually did. Because they adjusted for so many interacting variables, what the Drexel team did, in effect, was not much different than their main analysis of the entire population, as seen in Figure 3.4. They essentially took sickness and oxidative stress out of the equation in all of their analyses. This is equivalent to blaming the Cy2D6 protein for the death of an infant after the child is given codeine.

This tragically misleading and erroneous result is directly due to the *assumption* underlying the analysis. It is strictly an assumption, and again we know that the assumption is not valid, since oxidative stress, associated with the variables they treated as confounding factors, interacts with acetaminophen. The final data point on the far right of Figure 3.2 that includes Drexel's sibling analysis is not much different than the data point next to it, which is the analysis of the entire population after dozens of oxidative stress–related factors have been essentially canceled out. However, the sibling control analysis probably does cancel out some of the contribution of genetics. This is also not valid because genetics play a role in how acetaminophen is metabolized. Genetics are another variable that interacts with acetaminophen. It cannot be "assumed" to be a confounding factor.

To look at it another way, on the left-hand side of Figure 3.4, you can see that acetaminophen use during pregnancy is definitely associated with autism. What's causing the association is *either* the actual acetaminophen use, *or* something else that is somehow associated with acetaminophen use. The authors' conclusions were that it is something else. This conclusion is strictly dependent on an assumption used in the analysis that is known to be wrong. Abundant evidence described in this book tells us that the conclusion reached by Drexel scientists, and the assumption underlying that conclusion, are both wrong.

When looking at the results from the Drexel study, it's important to keep in mind that the authors are only looking at acetaminophen exposure during pregnancy. But based on available evidence,[109] pregnancy is probably *not* the time when most autism is induced by acetaminophen. For a sibling control analysis to work, the investigator would need a series of similarly sick sibling pairs in which one sibling was exposed to acetaminophen during a period of high risk for acetaminophen exposure (probably not pregnancy) and the other was not. For the most valid results, acetaminophen exposure would need to be somehow randomized, which is something no parent would reasonably agree to. Parents do what they believe is best for their child, and that is not random. However, acetaminophen is sometimes given during labor and delivery, and, as will be discussed in chapter 5, that time is likely the riskiest for acetaminophen exposure. Since acetaminophen does not effectively control pain during labor and delivery, whether the mother receives the drug is often more a matter of hospital policy or physician preference than actual need. In addition, the timing between acetaminophen administration and clamping of the umbilical cord—the moment the baby is left alone to process the drug—is highly variable. None of that is random, but it's mostly random with respect to the variables that we care about, which relate to the health of the mother and child. So, in theory, a sibling control study with lots of siblings comparing exposure during labor and delivery might work.

But such a study would be highly problematic if not impossible from a practical perspective. It's not ethical to expose women to acetaminophen during labor and delivery just for a scientific study, and it seems highly unlikely that enough samples of umbilical cord blood are currently stored in the freezers of our research institutions to conduct such a study. The main problem is that hundreds of samples must be collected and stored just to get a few samples connected with an autism diagnosis. We can't use medical records of acetaminophen use to determine exposure during labor and delivery because the impact of acetaminophen given before a very long labor is going to be very different than acetaminophen given just before a C-section or very short labor. We need to know what was going on at the instant the umbilical cord was clamped, which tells us how much acetaminophen the newborn had to metabolize without the mother's help. In addition, we would need siblings separated in time for our hypothetical study. Twins won't work because they would have the same exposure to acetaminophen, for practical intents and purposes, at birth. The long and short of it is that it is possible to *design* a valid sibling control study, but probably impossible to *execute* that study. Drexel executed an invalid sibling control study.

After the Drexel study came out, our research group worked some long hours to, as quickly as possible, demonstrate the problem with the Drexel study. We combined this new work with background and introductory information that had been previously rejected by the journal *Pediatrics*. (The events surrounding the rejection of our manuscript by *Pediatrics* are described in chapter 5.) This new manuscript, published months after the Drexel study,[110] was a computational study. In other words, we used computer-generated (simulated or virtual, not real) data in our analysis. When we set up the experiment, we defined exactly which virtual babies received acetaminophen, which had oxidative stress, and we defined the interactions between acetaminophen and oxidative stress that led to autism. Then we analyzed the *virtual* data the same way Drexel scientists analyzed *human* data. Some of the results are shown in Figure 3.5.

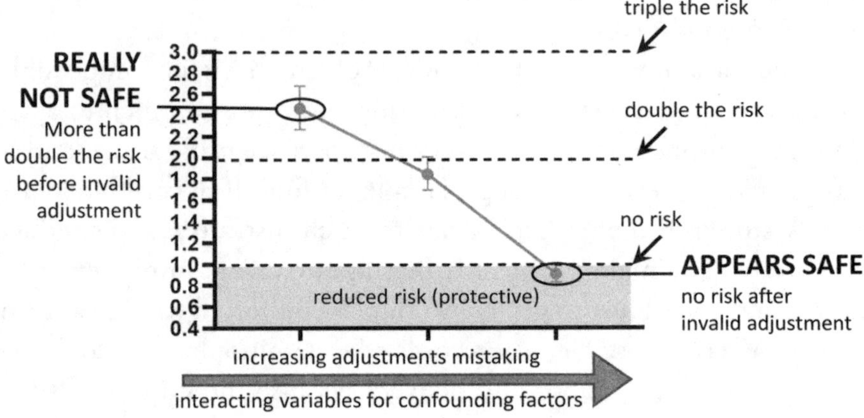

Figure 3.5: Some of the results of treating interacting variables as confounding factors in a simulated study published in *Life* in 2024.[111] In this study, we asked the question, what result would we get if half of all cases of autism were induced by exposure to acetaminophen during the simulated study period, in combination with oxidative stress? As the factors related to oxidative stress were eliminated using the same type of calculations used to assess data in humans, the obvious *real* risks of acetaminophen use, on the left side of the graph, were canceled out, as shown on the right side of the graph. This graph is reproduced from the same video described in Figure 3.4, which can be found on the publisher's website.[112]

It's evident from Figure 3.5 that we essentially duplicated the results from Drexel scientists using a simulated model system in which acetaminophen *did* cause half of all autism. That much is proven.[113] Thus, the calculations the Drexel team and others used yield the same result we would expect if indeed acetaminophen plus oxidative stress caused autism.

To summarize the situation,

- Our simulation showed the results to expect if acetaminophen plus oxidative stress caused autism and we assumed that factors related to oxidative stress were confounding factors.
- The Drexel study published in *JAMA* in 2024 got the result we'd expect if acetaminophen plus oxidative stress causes autism.

While this demonstrates that the Drexel team's calculations obscured a real association, it doesn't prove that acetaminophen plus oxidative stress causes autism. We need to look at more evidence to know if acetaminophen is involved in the induction of autism. What it does prove is that we can't draw the conclusion from the Drexel study that the authors drew and that most people are drawing. We can't conclude there is no association between autism and acetaminophen use during pregnancy, and therefore that acetaminophen can't cause autism when used during pregnancy. At this point, the scientific literature contains dozens of original studies and analyses of those original studies. Literally, *The BMJ* just published an analysis of the analyses of the original studies.[114] All of the original studies and the analyses of those studies, and the analyses of the analyses of those studies, contain the same, rudimentary error. Every single one of them is tragically harmful and misleading.

When considering the above discussion, one might wonder how anybody can draw any conclusions without addressing confounding factors that might exist. Shouldn't we try to adjust for *something*? The bottom line is that all the things associated with neurological problems, the things that might be considered confounding, are *also* associated with oxidative stress.[115] This includes genetics, all kinds of mental health problems, infections, smoking, autoimmune diseases, other medical conditions—as well as the medications used to treat those conditions, such as antibiotics—and even the number of children that a woman has borne. Even adjustment for socioeconomic factors isn't legitimate. Poverty and other socioeconomic factors have long been known to be associated with oxidative stress.[116] Our 2017 review,[117] the one coinciding with the beginning of my problems with administration at Duke University, covered this issue in some detail.

The bottom line here is that the associations that exist between acetaminophen use and autism are what they are. There is no legitimate way to explain them away with confounding factors. If we want to know something about whether those associations represent causal relationships, then we need to look at other evidence.

In 2020, when my group at Duke decided to address this rudimentary statistical error, I decided to talk to a statistician. I found a full professor of biostatistics and bioinformatics who had helped many Duke faculty with their projects. She had a wonderful reputation and worked with other Duke faculty as part of a service agreement between her department and the Department of Surgery, where I was located. I laid out the project for her in succinct terms: We want to run a simulation in which an independent variable (acetaminophen use) interacts with a lot of other factors (factors related to oxidative stress) that combine to cause a dependent outcome (autism). Then we will analyze the data by assuming that those other factors (related to oxidative stress) are confounding factors. As soon as I finished that description, she looked at me and, with no hesitation, said simply, "That won't work." I explained that this was what other people were doing, and we needed to *show* that it wouldn't work. She agreed, and our paper covering the topic was published in 2022.[118]

That paper didn't seem to make a difference as far as we could tell, so we concluded that more work was needed to push the issue further into the collective scientific consciousness. In addition, the aforementioned Drexel paper came out in 2024 in *JAMA*[119] and was even worse (more wrong) than most previous work on the topic. That provided the final push for us to publish another paper on the topic. I had retired from Duke by that time, so we didn't have access to Duke's statistical support, but we were able to use the calculations from Duke as a cornerstone for our work. We were able to repeat our previous results exactly, to the fifth decimal place, so we knew we were running the calculations consistently. Our second publication on the topic, in 2024,[120] covered the error in much greater detail and focused strictly on the error rather than covering a variety of topics related to acetaminophen and autism.

It's important to keep in mind the fact that, just from looking at the data from Drexel (Figure 3.4), there is no way to know whether those other variables are confounding factors or interacting variables. This is one of the limitations in the science of epidemiology. We must look at other fields of science to tell us whether the system is flooded with confounding factors or interacting variables. That is, of course, exactly what we did.[121]

At this point, after reading this far in the book, you can probably look at many scientific articles dealing with acetaminophen and autism, read the abstract, and see some of the mistakes. If they start talking about confounding factors, they are probably wrong. You'll also note that this "consensus" that acetaminophen is safe is not a true consensus. Multiple, independent individuals are not coming up with this conclusion. Rather, the problem is an "illusion of consensus."[122] What started as a rudimentary error in 2013 has spread without critical assessment.

You might be wondering why proof of fatal flaws that affects most studies evaluating the connection between acetaminophen and autism has not been widely publicized. We had what we thought was a good opportunity to change that—but a breach of scientific ethics prevented that from happening, as you'll see.

In mid-August 2025, a group at Harvard, led by Professor Andrea Baccarelli at Harvard's T.H. Chan School of Public Health, published a review of studies looking at associations between acetaminophen use during pregnancy and autism.[123] That review was the first to acknowledge our proof that treating interacting variables as confounding factors dramatically underestimated the impact of acetaminophen on the induction of autism. That review was highly influential, being touted by the Trump administration in their discussion of the connection between acetaminophen and autism. Unfortunately, that review referred to our proof of error, described in this chapter, as "speculative."[124] The authors then proceeded to ignore the error entirely, profoundly underestimating the association between acetaminophen use and the prevalence of autism.

In general, in the field of science, when an investigator disparages your work in the public format of a journal publication, journal editors give you the opportunity to rebut the negative view of your work. In other words, you are allowed to defend yourself. A number of years ago, I was involved in such a case.[125] In general, it's best not to disparage somebody else's work in the peer-reviewed literature unless you really know what you are talking about. They will get a chance to defend their work, and rightfully so.

In this case, Baccarelli had belittled our work in the journal *Environmental Health,* a peer-reviewed format, calling our mathematical proof "speculative." Our team dutifully wrote a short article that met the journal's guidelines for length, explaining that it was physically impossible for factors related to oxidative stress to be uninvolved in the toxicity of acetaminophen. In other words, susceptibility to injury by acetaminophen had to be real. It was not "speculative." This is based on the known principles of the science of pharmacology as they relate to acetaminophen metabolism as described in chapter 2. We submitted the article to *Environmental Health* on August 17, 2025, literally three days after Baccarelli's paper was published by the journal. The day after we submitted the rebuttal, on August 18, I was notified by the editor, Professor Philippe Grandjean, a founder of the journal and an adjunct professor at Harvard's T.H. Chan School of Public Health, that the paper would not be accepted. The rejection letter read as follows:

Your manuscript deals with adverse effects of a drug. While we agree that this issue deserves attention and abatement, we believe that your manuscript does not contain enough new environmental health information for our international readership and that your manuscript would be better placed elsewhere. Regrettably, your manuscript has been rejected for publication in *Environmental Health*.

We submitted the manuscript again, explaining to Grandjean that we were responding to a paper that *his journal* had published. Since his journal had already published a paper on the topic, and since our paper described substantial errors in that paper, it was not possible that our paper did "not contain enough new environmental health information for our international readership." But, again, we received a rejection letter, this one identical, word for word, to the first. We suspected that we were getting form letter rejections from an administrative assistant, but on August 27, Joliedianne Leobrera, "Journals Editorial Office Assistant" for the publisher, Springer Nature, assured us that Grandjean was indeed making the editorial decisions. The same day, we informed Leobrera of "irregularities" in the journal's review process. Three of the irregularities as outlined in our letter to Leobrera (text unedited except for an explanatory parenthetical insert and one typographical error correction) are as follows:

- Our work has been strongly and unjustly criticized by a paper published in his (Grandjean's) journal. It is customary for editors to give authors a chance to rebut strong criticisms that are published in their journals.
- The rationale for the rejection is verifiably false, or else the standards of the journal have been changed immediately prior to submission of our work. Since the journal published a paper on the topic only days before our submission, and since we are responding to that paper published in the journal, it is not possible that the journal does not address this topic. Further, since our paper refutes the conclusions raised in a paper the journal has published, it is not possible that we have contributed less to the field than a paper already published in the journal.
- Professor Grandjean and the senior author of the study in question are affiliated (or have recently been affiliated) with the same school. Normally in these cases the professor would excuse himself or herself from the decision-making process.

On September 23, 2025, the day after the Trump administration made their announcement about the connection between acetaminophen and autism, we received a final rejection of the manuscript addressing the fundamental and critical error in Baccarelli's paper. In this case, the letter came from Ruth Etzel, Professor at the George Washington University. Etzel and Grandjean have worked together for several years, for example on a paper describing factors that "hamper the translation of research findings into prevention of environmental hazards."[126] The letter from Etzel, with no reviewer comments, is as follows:

Dear Dr Parker,

Your manuscript "Underlying assumptions and a narrow focus by Prada et al result in futile recommendations" has now been assessed. If there are any reviewer comments on your manuscript, you can find them at the end of this email.

Regrettably, your manuscript has been rejected for publication in *Environmental Health*. Our journal aims at providing front-line research information to our international readership. We suggest that your Comment would be best addressed directly to the authors of the article in question.

Thank you for the opportunity to review your work. I'm sorry that we cannot be more positive on this occasion and hope you will not be deterred from submitting future work to *Environmental Health*.

Kind regards,
Ruth Etzel
Editor
Environmental Health

Given the high profile of Baccarelli's paper published in *Environmental Health*, it is indeed regrettable that a follow-up to the paper, explaining how the authors profoundly underestimated the association between acetaminophen and autism, was rejected. Yet another chance to educate the public was lost. It is ironic that the editors who rejected the paper, Etzel and Grandjean, had together written a paper lamenting unscrupulous efforts to prevent information about environmental toxins from helping the public.[127]

In the next chapters, we will dig deeper into the evidence connecting acetaminophen and autism. Fortunately, there's plenty of evidence to be found. It's not possible to induce autism in 3 percent of the entire population without leaving a massive trail of evidence.

CHAPTER 4

ASSOCIATIONS BETWEEN ACETAMINOPHEN AND AUTISM IN PLACE AND TIME

In chapter 2, I provided a very brief overview of information telling us that autism is not genetic, including evidence that autism was relatively rare prior to 1980, when acetaminophen replaced aspirin for children because aspirin was associated with Reye's syndrome. We also briefly looked at evidence indicating that autism is rare in parts of the world that lack modern medications. Those associations in place and time are important. As with the associations between acetaminophen use during pregnancy and autism described in chapter 3, associations between acetaminophen use and autism through place and time are not—by themselves—overly convincing. These associations do not allow us to draw any firm conclusions about causality. But if we knew that autism was common *before* acetaminophen was commonly given to babies and children, then nobody, including my research team, would likely consider the possibility that acetaminophen is important in the induction of autism. Here again, if an association does not exist, then, with rare exceptions, there can be no causal relationship. For that reason, associations in place and time are important.

Before we dig into the evidence in more depth, I will consider in more depth this simple and intuitive rule that, if association does not exist, causality can't exist. We recall that our burglar, Joe Banana, could not have broken into a jewelry store and stuffed his pockets full of jewels in

France if he was in New Mexico at the time of the burglary. However, this rule does have limitations. For example, if two different factors lead to the same outcome, then associations can be blurred even if causality exists. To explore this issue in more detail, we should consider the following rule, or equation:

Equation 4.1: acetaminophen exposure + susceptibility = autism

In our published work, we have concluded that the simplest explanation for all available observations is that this simple equation governs "the vast majority" of all cases of autism. We are certain without reasonable doubt that this equation governs "many if not most cases," but based on the principle of Occam's razor, which favors the simplest explanation, I favor the view that this rule governs the vast majority of all cases of autism. Here the phrase "favor the view" means that I am not certain that this equation governs the vast majority of all cases of autism, but rather that I think that it probably or most likely does govern the vast majority of cases. In other words, I give it better than a 50/50 chance compared to the alternative, which is that one or more things, in the absence of acetaminophen, can trigger a significant number of cases of autism. For the sake of discussion, let's define the vast majority as 11 out of every 12 cases, or about 92 percent. So, to summarize here, I think it likely (greater than a 50 percent chance, probably around 75 percent likelihood) that acetaminophen exposure triggers more than 90 percent of all cases of autism.

It's important to keep in mind that we aren't ignoring the fact that dozens of genetic, epigenetic and environmental factors are involved in the induction of autism. That's the susceptibility component of the equation.

But what if acetaminophen only triggers 30 percent of all cases of autism? In that case, the changing prevalence of other factors that cause autism may be even more important than changes in the amount of acetaminophen used. We might find that, in places without acetaminophen, or before 1980 when acetaminophen was rarely given to babies and children, the prevalence of autism was not much different than what we see today in the United States. Further, if acetaminophen triggers only 30 percent of all cases of autism, changes in diagnostic criteria, social awareness, economic factors and other issues could be more impactful on the prevalence of autism than changes in acetaminophen use.

What I've just said is that, even if acetaminophen induces many, say 30 percent, of all cases of autism, we might still find a high prevalence of autism in places and times with relatively low amounts of acetaminophen. However, based on our assessment, this does not appear to be the case. It appears that the prevalence of autism is very low without acetaminophen. This chapter will give you the evidence.

The first evidence we will examine in detail in this chapter involves the association of acetaminophen and autism with *place*. This evidence involves comparing the prevalence of autism in one part of the world with the prevalence of autism in another. Many readers will appreciate the fact that comparisons of autism between one place and another can be very difficult. Differences in cultures can make assessments challenging. Further, it can be extremely difficult to perform diagnostic evaluations in some cultures, as we will see later in this chapter when it comes to the Amish community. However, in general, we want to compare the prevalence of autism in places that have abundant modern medicine with places that have essentially no modern medicine. In doing so, we want to answer the questions, what is the prevalence of autism in areas without much or even any acetaminophen, and how does that compare with the amount of autism we see in the Western world? We will address this issue first and then come back to associations in *time*. As I address the evidence, I'll provide some examples of how we critically examine evidence and decide what is reliable enough to publish, and I'll give you some examples of the responses we've received to that evidence.

The best information we have from countries with very low levels of acetaminophen use is from Central America, as I pointed out in chapter 2. Here I'll provide a table that summarizes all the evidence we have included in our publications (Table 4.1). I'll list the evidence in order of how compelling we think the evidence is, with the most compelling evidence at the top.

Location	Observation
Leon, Nicaragua in 2010	No awareness of autism-related symptoms was found, even among educators, and a ratio of Down syndrome to autism of 33 to 1 in a school for special needs children was observed.[128] Based on current prevalence rates in the United States, the ratio of Down syndrome to autism is expected to be about 1 to 19 (1 in 691 or 0.14% with Down syndrome[129] and 1 in 36 or 2.8% with autism)[130] rather than 33 to 1. These numbers indicate that, using Down syndrome as a kind of "baseline," Nicaragua had a vastly lower prevalence of autism than the United States. Nicaragua has been plagued by shortages of even the most basic medications for decades.[131]
Six African countries in 1975 and 1976	"Marked autistic behaviour" was identified in only 5 of more than 400 (< 1.25%) "severely subnormal" children.[132] Between 5 percent and 8 percent of "severely subnormal" children were expected to have "marked autistic behaviour" based on studies in Britain.[133]
Somalia, prior to 2017	A sample of Somali immigrants living in North America viewed autism as a Western disease, with many believing that it does not exist in their native country.[134]
Cuba in 2016	Cuba may have a low prevalence of autism, with 241 cases of autism identified in the entire nation (1 in 25,000 children).[135] Acetaminophen is available in Cuba by prescription only,[136] and multiple travel advisors cite acetaminophen in particular as being in short supply in Cuba.[137]
Ethiopia, prior to 2004	Individuals with pervasive developmental disorders, about 50 percent of whom typically have autism, were found among second-generation Israeli immigrants from Ethiopia, but not among first-generation Israeli immigrants from Ethiopia (p = 0.0022).[138]

Table 4.1: Evidence indicating that the prevalence of autism is low in areas of the world with chronic shortages of medication. Only evidence that our team has included in our publications is shown. The notation "p = 0.0022" indicates that, roughly, the chances that the observations were due to random happenstance are about 2 in 1,000.

The evidence in Table 4.1 is "very compelling" in our view. It *could* be explained away, but it's not trivial to do so. One reporter correctly pointed out that, in the data from Nicaragua, they had no way of assessing the presence of mild cases of autism. What the reporter is suggesting is that, in Nicaragua in 2010, for some unknown reason, they might have had many cases of mild autism that we could not detect, despite having very few cases of more severe autism. This hypothetical condition brings up a series of concerning objections and seems unlikely. More important, this hypothetical scenario doesn't avoid the association between a lack of *severe* autism and a lack of acetaminophen we see in Nicaragua in 2010. Again, I'll remind the reader and myself that the evidence in Table 4.1 is going to fit beautifully with other lines of evidence, including associations between acetaminophen use with autism in time that will be covered later in this chapter. None of the single lines of evidence I'm presenting is meant to stand alone for the purpose of drawing conclusions. Also, please note that we consider the consolidation of all the items in Table 4.1 as a *single* line of evidence.

Some members of the news media have written articles which could be seen as an attempt to "debunk" the view that autism is rare if the use of acetaminophen is rare. An example of such work can be found in an article by Mary Kekatos published by ABC News on October 14, 2025 ("Debunking 3 Claims About Tylenol After White House Links Drug's Use in Pregnancy to Autism"). Kekatos graduated from Sarah Lawrence College in 2015, where she concentrated in journalism and twentieth-century European history. She then finished graduate school at Columbia University's Graduate School of Journalism the next year, in 2016. Then, less than a decade from graduation, she found herself writing an article "debunking" claims about acetaminophen and autism, one of the most important medical issues of our time.

At this point, I want to emphasize that I understand that the media is in a very, very difficult position. They are mostly journalists, not scientists. Kekatos went to a highly reputable source, the American Association of Pediatrics (AAP), for much of her information. As of October 2025, the Association was still claiming that "acetaminophen is safe for children when taken as directed" and that there is "no link to autism."[139] Essentially, Kekatos reported what she found, which was that the world's most prestigious pediatric association thinks that acetaminophen is safe for children. Of course, the opinion of the AAP does not "debunk" the idea that acetaminophen induces autism any more than the Catholic church's beliefs in the seventeenth century "debunked" Galileo Galilei's knowledge that the Earth revolves around the sun.

The point here is that Kekatos is not an "investigative reporter," and she does not pretend to be. It is not her job to learn all the relevant science and parse it out for her readers. It is only her job to find the consensus and to report that consensus. If Kekatos's report were a scientific report, she would be very guilty of a classic logical fallacy, called an *appeal to authority*, and her work would be immediately rejected as nonscientific. Her report is what it is. It's a review of what the recognized authorities believe to be true. I don't disagree with the fact that the AAP is not yet enlightened regarding the role of acetaminophen in the induction of autism. In my experience, the AAP has some gatekeepers that need to either abandon their dogma in favor of scientific evidence or they need to move out of the way. I'll cover that more in a later chapter. But for now, why do I bring up Kekatos's report here?

One of the things that Kekatos's report "debunked" was a claim that "Amish children don't have autism because they don't take acetaminophen." More reasonably, I would say that "The Amish may have lower levels of autism due to lower levels of acetaminophen use." Journalists sometimes take license to be dramatic, but here I'll consider the questions using less dramatic, more realistic language: Do the Amish have less autism, and do they use less acetaminophen? You'll notice that the Amish are not on our list of evidence shown in Table 4.1. We have, for years, considered data from the Amish. My team determined that the evidence obtained from the Amish is simply too weak at the present time to add to our publications on the topic. Shortly before the connection between acetaminophen and autism was announced by the US president, a newer member of our research team brought up the idea of including the Amish in our list of evidence. The idea was new to her. I'll show you what I wrote to her by email on July 30, 2025. I've fixed a couple of typographical errors, deleted personal information, and added parenthetical inserts for clarity.

Scientists have tried to get systematic access to the Amish community. My group even had access to funding for (a study of autism prevalence in the Amish community). But I understand that every time somebody tries to do a systematic screen of Amish children in a church or school, they get shut down by church elders.

The concern with the survey approach used in the study by Robinson et al is that (a) they probably only get consent from parents that are (less averse toward interactions with members outside their community, and who may also adhere less to other Amish traditions), and (b) they may be less likely to get consent from parents who have a child with special needs.

My guess would be that the Swartzentruber Amish have even a lower prevalence of (autism) than the Old Order Amish in general. They are generally stricter than average.

In the end, if we did have good data on the Amish, it would go under the subheading about lack of medicine leading to lack of autism. Our best data on that topic are from the ratio of Down syndrome to (autism) in Central America.

These are the sorts of deliberations that occur within our group on a regular basis when we determine what evidence is reliable enough to present in our scientific work. In the case of Cuba, which is now included in our table of published evidence (Table 4.1), we debated about the topic for years before we published it. We had incredible difficulty confirming the prevalence of acetaminophen use, until we found multiple reports on travel sites indicating that the country has chronic shortages of acetaminophen. On the other side of the coin, we had included some evidence from South Korea in some of our earlier work, but we now omit that evidence because we no longer think it's reliable enough. In the current example with the Amish, we had long been aware of the "Robinson et al." report, which was a survey study published in 2010[140] showing relatively low levels of autism, but it simply didn't make the cut due to concerns about reliability. So, from our point of view, information from South Korea and from the Amish go into the pile of "something fishy related to acetaminophen use might be going on there, but we aren't going to describe that as evidence because we just aren't certain enough about it."

In her report published by ABC News, Kekatos did a great job of uncovering what our research team had previously concluded. The experts Kekatos interviewed were really on top of that data from the Amish community and saw that it was unreliable. We agree that the data from the Amish really isn't strong enough to be compelling. That being said, we consider all the data shown in Table 4.1, when combined, to be compelling. Unfortunately, Kekatos's report provides no evidence that any of her experts knew any compelling data. Tragically, her report in a prominent and respected media outlet may lead some to conclude that acetaminophen is safe for babies and children.

I'll quickly point out here that it's not surprising that Kekatos didn't find experts with extensive knowledge of evidence connecting acetaminophen with autism. Countless hundreds or perhaps even thousands of scientists are capable of spotting some of the major limitations in that report on the prevalence of autism in the Amish. But only investigators who have spent thousands of hours *specifically* probing the connection between acetaminophen and autism will know what evidence exists and where that evidence is located. Knowing how to critique a paper is part of all scientific training, but knowing a scientific subject takes years. I would estimate that it takes about ten thousand hours to be competent, which is probably about the amount of time spent earning a PhD. It probably takes about twenty thousand hours to be an expert on a scientific subject, which means adding on a postdoctoral fellowship and a couple of years as a junior faculty after a PhD. Or, in my case, working on the subject as a faculty member for a decade.

It's important to keep in mind here that clinical practice does not count as scientific study. Using those criteria, Kekatos's report does not indicate that she consulted an expert on the subject she was covering. The report doesn't even indicate that she consulted anybody who was competent. None of the people she consulted realized that the best evidence regarding a low prevalence of autism in the absence of acetaminophen has nothing to do with the Amish.

In our current published list of evidence related to associations in place, we include some information from Scandinavian countries. This evidence is a bit different from the evidence shown in Table 4.1, which examines the prevalence of autism in areas of the world with very low levels of acetaminophen use. Evidence from Scandinavian countries compares the amount of acetaminophen used in Denmark and in Finland with the amount of autism in those countries from 2006 through 2010. We first published that evidence in 2024,[141] and I'll include the graph here:

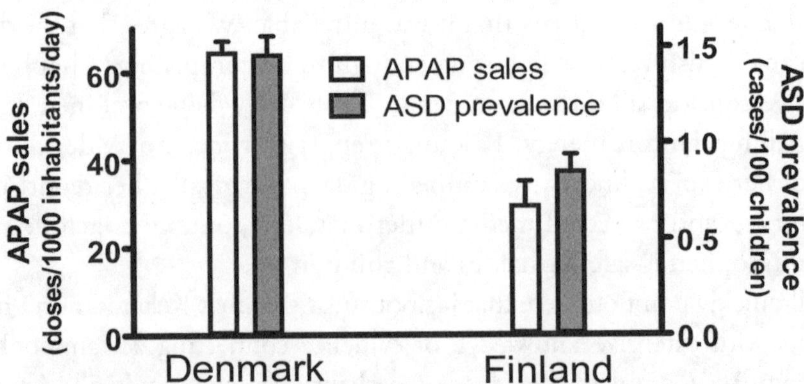

Figure 4.1: The sales of acetaminophen (APAP) in Denmark and Finland correlated with the prevalence of autism spectrum disorder (ASD) in those countries. The graph is published in the peer-reviewed literature. (This information is discussed in more detail in Chapter 7; see also Figure 7.3.).[142]

As scientists, when we decide to publish data of this sort, we are very keenly aware that association is not the same as causation, and that a dozen different factors might be responsible for the association shown in the graph. At the same time, we also know that the association shown in the graph is consistent with the view that exposure of susceptible children to acetaminophen induces many if not most cases of autism. As we stated in the paper, "When considered in light of other lines of evidence, this association adds to the burden of evidence demonstrating without reasonable doubt that acetaminophen use in susceptible babies and children causes many, if not most, cases of ASD (autism)."[143]

Despite the disclaimers we included in the paper, we still got lectures from "experts on social media" about associations not equaling causation. We also got vitriolic complaints about the publisher of the journal, claiming that only a horrible journal would let such nonsense be published. We published the paper with a journal called *Children*, which is owned by the publisher MDPI. In my experience, MDPI does a good job of getting reviews done quickly. We don't always get reviews that we like from MDPI, and I'll show you a review from an MDPI journal later in this book that is shocking, and not in a good way. Whether you encounter biased reviewers while working with MDPI journals depends more on the editor than on MDPI, but that's not the point here. The point is that they tend to complete the review process in a timely fashion.

A timely completion of the review process is important because the work is completely on hold while the paper is under review. Within the past year, my team waited for over six months for the editors working for one publisher (PLOS) to review one of our papers, and over nine months for the editors working for another publisher (Wiley) to review another of our papers. In both cases, we had to start over after months of waiting, with none of the editors providing any intelligible rationale for the rejection of our work. I will describe the situation with *PLOS One* in more detail in chapter 6. Both journals were reviewing papers dealing with the connection between acetaminophen and autism. Fortunately, both of those papers are available to the public,[144] but the point is that some publishers can take a very long time. With MDPI, the review process can often be completed within two or three months, which is reasonable and sometimes very important.

Some MDPI journals also have great readership, which can be very useful if the science has important implications for public health. The paper we published[145] with the evidence from Scandinavia (Figure 4.1) has been viewed more than 54,000 times, which is unusually high for a science paper. More importantly for this discussion, the journal we selected includes the academic editors' names with the publication. In other words, the two editors that approved our paper, Marco Fiore, a scientist at the National Research Council of Italy's Institute of Biochemistry and Cell Biology, and Luigi Tarani, a professor and pediatrician at the Sapienza University of Rome, attached their names to the paper.[146] It is a sign of a quality journal when the academic editors are identified, an indication that complaints about our journal selection are unjustified. More importantly, attacking evidence by pointing out perceived flaws with the publisher delivering the evidence constitutes the well-known logical fallacy of attacking the messenger, the ad hominem fallacy.

In this next section of this chapter, I will outline associations between autism and the use of acetaminophen through time. Several lines of evidence listed in chapter 2 and Appendix B involve associations between acetaminophen use through time. These include the increasing prevalence of autism after 1980, when acetaminophen replaced aspirin as the go-to drug for pediatric use. Another line of evidence related to time is the increasing prevalence of autism in the 1990s, when direct-to-consumer advertising was introduced and marketing of over-the-counter medications went ballistic. Finally, an examination of the writing of investigators during the 1980s and 1990s is worthwhile. Those investigators had a firsthand view of the changing prevalence of autism, and their writings are informative.

As my group and others have outlined many times, the beginning of the autism pandemic began when modern pediatric medicine switched from aspirin to acetaminophen in 1982–1983. Randy Bollinger, one of my friends and colleagues who's worked with me for decades, remembers when the switch was made. At the time, he explained to me that they had two patients in the hospital at Duke with Reye's syndrome. Since aspirin was suspected of triggering Reye's syndrome, which occurred in about one in ten thousand individuals at that time, pediatricians switched to acetaminophen. Although the switch to acetaminophen began in about 1982, it would have affected children born in 1980. Those children born in 1980 could have been exposed to acetaminophen in 1982, while they were still young enough to be susceptible to the induction of autism.

Our research team has also connected direct-to-consumer advertising and marketing of over-the-counter drugs with the rise in autism. The prevalence of autism didn't stop rising after the switch from aspirin to acetaminophen was complete in pediatric medicine. The 1990s saw a dramatic rise in all medication use, with billions of advertising dollars spent promoting acetaminophen use. As we pointed out in the scientific literature,[147] between $115 and $250 million ($206–488 million /year with inflation adjustment for 2025) in direct-to-consumer advertising per year encouraged acetaminophen use.[148] In a single year, that's more than double, in some cases, the advertising dollars spent on a major Hollywood movie. The temporal association between direct-to-consumer advertising spending and autism prevalence is depicted in the graph in Figure 4.2, which was published in the flagship journal for the Korean Pediatric Society.

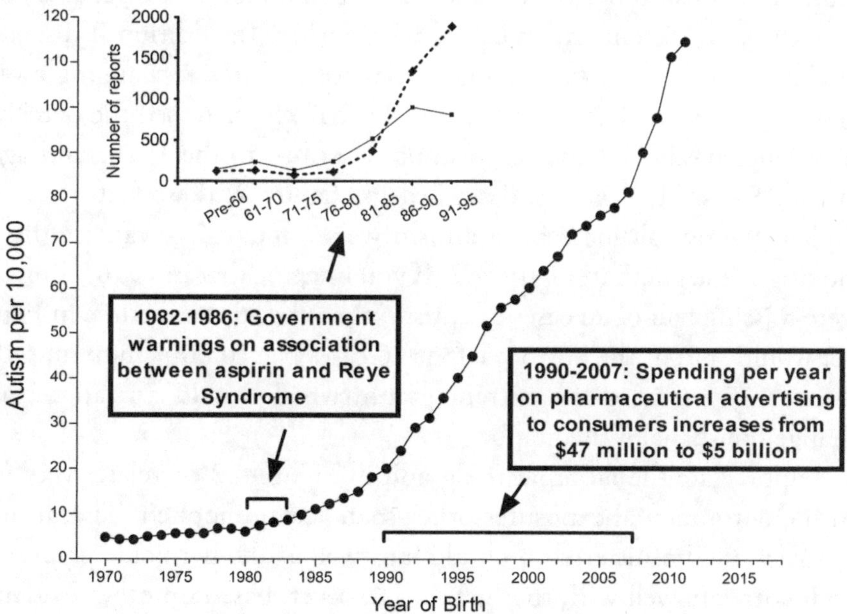

Figure 4.2: Temporal associations between the reported incidence of autism in California and factors affecting the use of acetaminophen. The prevalence of autism in California as compiled by Cynthia Nevison[149] is shown in the main graph. In the inset, previously published survey data[150] from the Autism Research Institute and the Autism Society of America are shown. In the inset, the number of surveys (reports) that were collected within a given time frame are shown, and reports are separated into those describing infantile (non-regressive or early-onset) autism (solid line) and those describing regressive autism (dashed line). This graph and portions of this figure caption were previously published in the peer-reviewed literature.[151]

As you might imagine, my research group has spent a considerable amount of time examining the graph in Figure 4.2. We know key events in history, such as the switch from aspirin to acetaminophen and the increase in pharmaceutical advertising, that undoubtedly affected acetaminophen use. Those are indicated on the graph. But nobody actually tracked acetaminophen use in the pediatric population. We've checked using systematic scientific approaches, so we should have found the information if it existed in the scientific literature.[152] The bottom line is that we just don't have accurate data on the pediatric use of acetaminophen through time. Apparently, the drug was considered so safe that nobody bothered to see how much was being given to babies and small children. In another chapter, we will address the error made in the 1960s and 1970s that first allowed acetaminophen into pediatric practice.

The graph in Figure 4.2 doesn't allow us to draw conclusions about the causal relationship between acetaminophen and autism, of course. But, again, it is consistent with a causal relationship. In addition, I discussed in chapter 2 the fact that changing diagnostic criteria, changing awareness, and social and economic factors have likely increased the perceived prevalence of autism since it was first described in the 1920s and again in the 1940s. However, as described in chapter 2, these factors do not explain the overall increase in autism we see today. We can see this by looking at the graph in Figure 4.2. If you imagine that in 1970, clinicians were missing half of all cases of autism, then the adjusted curve in Figure 4.2 would start at the bottom left at 10 cases per 10,000 children, rather than the 5 cases on the current graph, which would not appreciably change the shape of the graph.

I appreciate the fact that the graph in Figure 4.2 correlates very well with environmental exposures other than acetaminophen. The number of vaccines administered to children as well as the herbicide glyphosate both correlate well with that graph. However, based on other evidence, we can rule out vaccines and glyphosate as *specific* triggers for autism, and that will be discussed in another chapter. But what we can't rule out, and what we should probably consider seriously, is that the number of environmental exposures that contribute to susceptibility may be increasing. Here I will remind us of equation 4.1.

Equation 4.1: acetaminophen exposure + susceptibility = autism

Based on this equation, increases in factors that impart susceptibility will result in an increase in the prevalence of autism—even if acetaminophen use does not increase. At the same time, increases in susceptibility, which entails increasing oxidative stress, will likely correspond to increasing amounts of sickness, which will likely result in more acetaminophen use. Indeed, investigators at UCLA and Brigham and Women's Hospital in Boston reported that the prevalence of chronic conditions in children age five to seventeen years increased from about 20 percent in 1999 to more than 27 percent by 2018.[153] In other words, based on one assessment, the number of children affected by chronic conditions increased from twenty to twenty-seven out of every hundred children in less than twenty years. With that in mind, it seems possible that current increases in autism are induced at least in part by ever-increasing oxidative stress and sickness that corresponds to increasing susceptibility, which in turn results in increasing use of acetaminophen.

Another factor that tells us the prevalence of autism is actually increasing is the scientific literature itself. When scientists take measurements, they are usually quite aware of changes that might affect their measurement. Such changes will, of course, affect the ability of the scientist to compare data obtained in the past with data that will be obtained in the future. Given that investigators try to keep factors that affect their measurements under control, it is not surprising that the increasing prevalence of autism over time caused some degree of consternation. For example, in 1987 Matsuishi et al.[154] stated:

What is the reason for the high [1/645] prevalence rates? We doubt that differences in diagnostic criteria and survey methods account for these high rates, although such factors may explain the large differences between the studies done between 1966 and 1980 and the studies done since 1983. In fact, the earlier studies were carefully performed using good survey methods. Although in some studies the diagnostic criteria may have been different, the diagnostic criteria were similar in the study by Hoshino et al (reference) and in our study, and both studies were done in Japan in similar cultural, ethnic, and socioeconomic regions. However our prevalence is three times that of Hoshino et al. Could the real prevalence of autism be increasing in the world?

Almost thirty years later, in 2006, Williams and colleagues[155] noted that changes in diagnostic criteria did not clearly explain the rising prevalence of autism:

> . . . *it was not possible to account entirely for the effect of the diagnostic criteria on the prevalence estimates as the ICD-10 and DSM-IV diagnostic schema leave some scope for variation in their interpretation and application.*

Again, as Coury and Nash noted in 2003:[156]

> *If MMR and thimerosal* [factors related to vaccination] *aren't responsible, what's the causative factor for autism? The simple answer is that it is not known.*

When a scientist obtains a result that is different from that of other scientists, it is customary to attribute that difference to some factor related to the methods used for measurement. But when the same scientist gets different results at different times, then the situation is more concerning, and we tend to point the blame toward the thing we are measuring. For example, in my current laboratory, we found dramatic changes in rat behavior following exposure to acetaminophen, but then we obtained a completely different, yet equally concerning, result when we exposed a second group of rats to the same amount of acetaminophen.[157] In cases like this involving laboratory animals, scientists call these differences in results *cohort effects*, which means that something probably changed between the two sets of animals. Since we use laboratory rats that are not all genetically identical, a likely suspect is the genetics of the rats. By the same token, when it comes to assessing the prevalence of autism, scientists seeing a change in measured prevalence will tend to point to the population being measured as being the source of the change. If they didn't change something with their measurement, then the population itself must be changing. It makes sense.

In the next chapter, we will dig into more evidence, this time looking at evidence that tells us *when* the brain is susceptible to injury by acetaminophen. Specifically, we want to look at why labor and delivery is so important, and why pregnancy may also be important, but less important than the time of birth and the first couple of years of life.

CHAPTER 5

TIMING OF RISK: PREGNANCY, LABOR AND DELIVERY, AND EARLY CHILDHOOD

Risk Before and After Clamping the Umbilical Cord: No Contest.
The first scientific error described in this book was the confusion of interacting variables with confounding factors. That error was described in chapter 3. This chapter will describe a second error, focusing on a time period when acetaminophen is probably less dangerous than other times. This error is equivalent to using a study in fourteen- to eighteen-year-olds to prove that children playing with matches is not very dangerous. If the youngest "child" in the study was fourteen years old, then the results cannot be applied to children in general, as the large majority are under fourteen. Regardless of the outcome of a study looking at the outcomes of people *over* the age of fourteen—often designated "children" in the medical literature—playing with matches, a study of three-year-olds playing with matches would probably show very poor outcomes. If scientists look at the wrong age because they make assumptions about the timing of the risk, they can miss the big picture.

One of our conclusions stated in the introduction is that the most dangerous time for exposure to acetaminophen is probably immediately after birth, and extending to about one or two months of age. Early childhood, up to about three years of age, is probably the second-riskiest time. Some risk is still present at four and five years of age, and some risk is probably present during pregnancy with heavy use of acetaminophen.

A diagram showing what we think about the age of the brain and the risk of acetaminophen-induced autism is shown in Figure 5.1.

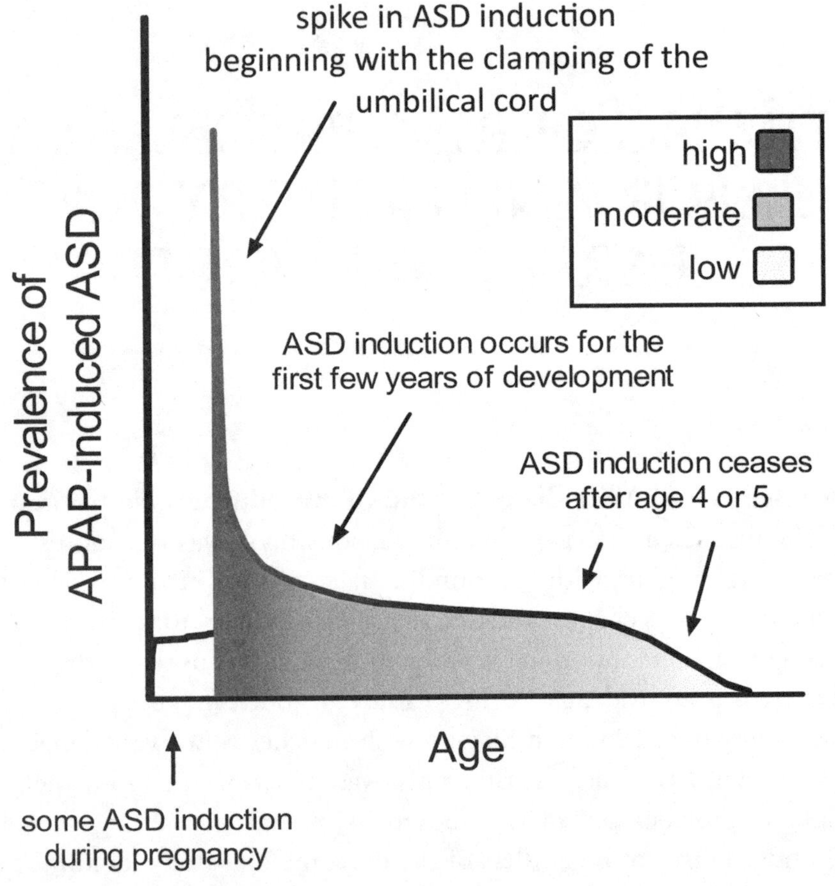

Figure 5.1: In this schematic diagram, the risk of a reaction to acetaminophen leading to autism increases dramatically when the umbilical cord is cut, then drops off significantly after a few weeks to months, slowly decreasing, with some risk at ages 4 and 5, but probably negligible risk at about age 6. A version of this graph was published in the journal *Children*.[158] Being a "schematic diagram," the graph represents our understanding based on available evidence, and does not show actual data collected from an experiment or study.

This chapter is going to explain how we arrived at the conclusions depicted in the graph shown in Figure 5.1. It's impossible to do a single experiment that would show us that graph. As with many things in science, one single experiment cannot tell us what is going on. We can't see the electrons move around the nucleus of a single atom, or a single protein fold into a wonderful, functional shape. But if we have enough information, we can deduce how the system works.

When coming up with the curve shown in Figure 5.1, we used three types of evidence that contribute to our understanding of the system.

- Evidence related to the connection between acetaminophen and autism at different times of life.
- Evidence related to the connection between acetaminophen and autism unrelated to a specific time of life.
- General scientific principles, including basic knowledge related to autism.

A diagram showing how various scientific observations led us to deduce the timing of autism induction (shown in Figure 5.1) is shown in Figure 5.2. Out of about thirty lines of evidence telling us that acetaminophen is important in the induction of autism, roughly eight of those lines of evidence can be used to deduce *when* acetaminophen induces autism. Several general scientific considerations, including the previously discussed Occam's razor, have a bearing on the conclusions.

The evidence shown in Figure 5.2 forms a coherent picture. All of the associations add up nicely to close to 100 percent of all autism being induced by acetaminophen.

- About half of all autism is induced at birth or within the first few weeks of birth.
- About 30–40 percent of autism is induced after months of development, leading to regression into autism.
- Less than 20 percent of autism is probably induced during pregnancy.

We are reasonably confident that the risk is greater immediately after birth than at other times. The strength of the associations match up with what we expect from known interactions between acetaminophen and humans at various ages. In other words, when we expect the drug to be most dangerous—based on what we know about age and ability to handle drugs—we see stronger associations between drug use and autism. It all makes sense.

If, hypothetically, we lined up a million randomly selected people, starting with the person most able to safely metabolize acetaminophen and ending with the person least likely to safely metabolize it, we would see a very dramatic pattern. At the front of the line, we would see a lot

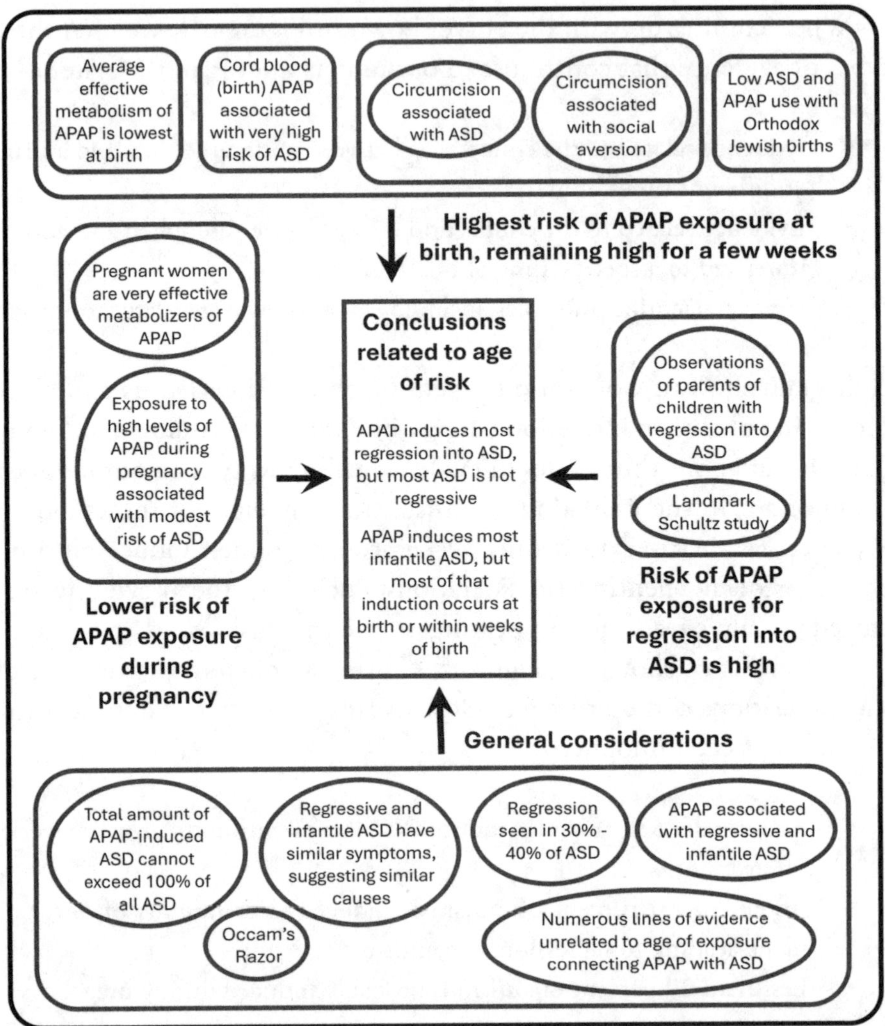

Figure 5.2: Evidence coming together to form a coherent picture of the connection between age and the induction of autism (ASD) by acetaminophen (APAP). Evidence and conclusions related to the age of risk are shown.

of pregnant women. At the end of the line, we would find people with autism and newborn babies. Pregnant women are great at metabolizing acetaminophen,[159] whereas newborns and people with autism are the opposite (see chapter 2).

Before we move on, it's important to address a question critical to this chapter: *When exactly does the risk of acetaminophen use begin to drop after birth?* When we talk about the time right after birth, we don't mean to suggest that risk decreases quickly within a few hours, or even a few days. Athena Zuppa's work at the Children's Hospital of Philadelphia suggest that a baby's ability to metabolize acetaminophen begins to improve about two months after birth.[160] Then, children's ability to metabolize acetaminophen steadily improves until about twenty-four months of age, when they metabolize acetaminophen as well as adults.[161] But Zuppa's team only had a few babies in their study younger than two months of age, so we can't be certain exactly when babies develop their ability to metabolize acetaminophen. In other words, we can tell from Zuppa's work that babies can't metabolize acetaminophen well when they are born. We can also tell that some toddlers at two months of age are better able to process acetaminophen than newborns. But we don't have a good idea of when exactly the average baby starts to metabolize acetaminophen better than a newborn. In addition, we know that some babies metabolize it better than others. Not every baby is the same.

Zuppa's work tells us with confidence that the average baby has a significant ability to metabolize acetaminophen by twelve months of age, and that a toddler's ability to metabolize acetaminophen, on average, is fully developed by twenty-four months of age. But that's only the *average* baby. Zuppa's findings don't tell us that *all* babies are safe by twelve months of age, or even twenty-four months of age. It's important to keep in mind that many children with autism can't clear acetaminophen using safe pathways that most children have. They are particularly deficient in sulfation,[162] one of the safe pathways, as noted in chapter 2. And children with autism are apparently also deficient in the other safe pathway, called glucuronidation, compared to other children, even when they are several years old.[163]

Let's take a more detailed look at where exactly we came up with the shape of the graph in Figure 5.1. First, how did we estimate that about 50 percent of all cases of autism are induced at the time of birth or within a few weeks of birth? That number comes from the Johns Hopkins study showing that the third of women with the highest amount of acetaminophen (and/or some of the chemicals that come from acetaminophen) at the time of birth have an almost fourfold increased risk of having a child with autism.[164] The problem here is not that the *women* have the acetaminophen, but that, practically speaking, the newborn babies get the same amount of acetaminophen—and the newborns have to process the drug all by themselves, without their mothers' help. If we run a very rough calculation based on the data from the Johns Hopkins study, we come up with 56 percent of all cases of autism are induced at the time of birth. I will emphasize that this number is based on a very rough calculation,[165] definitely not an exact calculation, but it's as close as we can get with the information we have. The number of cases of autism induced later in life, leading to regression into autism, is based on the numbers of cases that involve regression. That could be somewhere between 30 or 40 percent of cases.[166] But, again, the exact number of cases involving regression is not precisely known, and the number varies from study to study.[167] Later in this chapter we will discuss in more detail why we think that most cases of regression are induced by acetaminophen.

Finally, after considering risks in the first weeks after birth and then the first years of life, we come to risks during pregnancy. The best information, with the most data and the finest detail in the data, are still from Sweden, published by Drexel University.[168] For our purposes, it doesn't matter that the conclusions of that study were entirely wrong,[169] as described in chapter 3; those data are the best. Their data show an association between autism and acetaminophen use during pregnancy that is not nearly as high as the association with acetaminophen use at birth. Even large amounts of acetaminophen during pregnancy were associated with less than a twofold increased risk in autism compared to an almost fourfold increased risk at labor and delivery. In addition, not very many women take so much acetaminophen during pregnancy. (We'll go over that in more detail in the next paragraph.) A much greater percentage, one third of all women in the Johns Hopkins study, had enough acetaminophen at the time of birth to increase their risk of having a child with autism by three- or fourfold.

The Drexel team thought that less that 2 percent of women, about one in every fifty women, used enough acetaminophen to be associated with almost a twofold increased risk of having a child with autism. This number was probably an underestimate[170] because of the way the data were collected,[171] but heavy use of acetaminophen typically involves treatment of chronic pain, which probably affects about 10 percent, more or less, of the population of childbearing age.[172] So, the bottom line with induction of autism during pregnancy is that, mostly likely, few women use enough acetaminophen to be associated with an almost twofold increased risk of autism. Even for the women that do use that much acetaminophen, the risks are not as high as at the time of birth. Before the Drexel study came out, we previously estimated that the proportion of cases of autism induced during pregnancy was possibly less than 10 percent and probably not more than 20 percent of the total.[173] The Drexel study doesn't add any information that would cause us to increase our estimate of the percent of autism cases induced by acetaminophen use during pregnancy.

After all of that, you might wonder why scientists have been so focused on pregnancy. One reason could be that data describing acetaminophen use during pregnancy are easier to obtain and more reliable than data describing acetaminophen use at the time of birth and afterward.

If we combine the amount of acetaminophen-induced autism during pregnancy, the period of time after birth, and early childhood, it is nice to see that the numbers add up to about 100 percent of all cases of autism. This satisfies Occam's razor, providing a straightforward explanation for the induction of the vast majority of all cases of autism. But really, there is so much wiggle room in those numbers that we can't be sure exactly how many cases of autism are induced by acetaminophen. At least not from those numbers. On the other hand, other lines of evidence support the view that acetaminophen is involved in the vast majority of all cases of autism. The alternative explanation, that more than one factor can lead to autism, is less satisfactory from a scientific perspective. Again, we are keeping in mind that multiple factors cause susceptibility to acetaminophen. We do not think that acetaminophen alone induces autism. That idea is absolutely wrong. But we are thinking that acetaminophen is *essential* in the induction of that vast majority of all cases of autism.

We have identified no observations that stick out, leading to nagging questions. In other words, we are not aware of any observations that don't fit the conclusion that acetaminophen is essential in the induction of the vast majority of all cases of autism. This is not to say that we know all the details. At this point, for example, we know that acetaminophen forms increasing amounts of a toxic substance in the presence of genetic and environmental factors associated with autism. But we don't know what exactly that toxic substance does. The best guess is that it binds directly to proteins in the brain, particularly to proteins in the mitochondria, the "power plant" used by most cells in our body, including brain cells. In this way, the toxic substance formed by acetaminophen could critically injure brain cells, eventually leading to their death. But other mechanisms are possible; for example, the toxic substance formed by acetaminophen might affect the blood or the liver, and damage to the blood or the liver might indirectly affect the brain. That is one of the details that nobody has studied.

In general, my research team is not overly concerned about the details that remain unknown. This isn't a science project just for the sake of knowledge, and we know far more than enough to make adjustments to clinical practice. But not knowing all the details, or in some cases simply not having absolute proof about some of the details, can be used as justification to block progress despite overwhelming evidence. The view that "more research needs to be done" is ingrained in the scientific consciousness. That's hard to argue with . . . unless we have enough evidence to move forward on a topic of profound importance to public health. Then, we argue that action should be taken. In those cases, conducting more research is fine, but it's not fine to *wait* for more research. Indeed, taking action is the ultimate experiment. We expect the prevalence of autism to plummet when acetaminophen is avoided during early brain development, until the age of six years, when risk is apparently very low.

The typical reader of this book will probably eventually conclude, if they have not already, that we know more than enough to do what it takes to reduce acetaminophen exposure during brain development. However, some experts have disagreed. In general, they argue that some details regarding the interaction between acetaminophen and babies are unknown, and therefore we need more work before we can make any clinical changes. This argument in a general form is not acceptable. We will *never* know everything, which means that we can never take action if we must know everything before taking action. For the argument to be reasonable, an unknown factor must cast very substantial doubts on the conclusions.

One example of highlighting unknown details regarding the connection between acetaminophen and autism in an attempt to block progress comes from the renowned journal *Pediatrics*. *Pediatrics*, in publication since 1948, is the flagship journal of the American Academy of Pediatrics. The journal is highly influential, cited by clinicians and scientists more than any other journal in the field of pediatrics. Indeed, as you will see in chapter 8, that journal played a key role in accepting acetaminophen for pediatric use in the first place, before it was ever tested for neurodevelopment. My interactions with *Pediatrics* provide many useful examples of "nonscientific objections" to our work and will be considered again in chapter 9.

On February 7, 2024, I submitted a manuscript to *Pediatrics* (Manuscript number PEDIATRICS/2024/066146) describing our conclusions regarding the role of acetaminophen in the induction of autism.

Professor Alex Kemper, at the Ohio State University College of Medicine, was deputy editor of *Pediatrics* at that time and handled the manuscript. On February 12, 2024, Kemper informed us that the journal would only consider a very shortened version of the paper, called a *perspective*. We dramatically shortened the manuscript based on the formatting requirements for a perspective, which apparently antagonized the reviewers who eventually examined the manuscript. The shortened manuscript was submitted to the editor on February 17, 2024 (Manuscript number PEDIATRICS/2024/066146). The manuscript received two very negative reviews that were devoid of any valid rationale for rejecting its conclusions. Nonetheless, the manuscript was rejected on March 18 based on those reviews. We wrote a rebuttal letter containing all of the reviewers' comments and explaining how each one was invalid. Appendix C contains that entire letter. As a result of that letter, Kemper obtained another reviewer for the manuscript, and on April 10, 2024, we were given another review devoid of any rational concerns—and another rejection. We wrote yet another letter in response, this one shown in its entirety in Appendix D. The letter was presented to Kemper on April 11, and he informed us within minutes that the manuscript would not be reconsidered. In total, the reviewers presented about fourteen unique objections, the most popular of which was "you are wrong," without considering all the evidence. The paper was eventually expanded and published in the journal *Life*.[174]

In the second round of review (Appendix D), the reviewer points out that nobody has ever proven that the toxic metabolite of acetaminophen is produced more in patients with a variety of conditions involving oxidative stress.

> "Regarding factors associated with ASD that can cause oxidative stress—I have not seen evidence that sex, Down syndrome, preterm birth, high vitamin B levels, leads to elevated NAPQI production . . ." (anonymous reviewer, *Pediatrics*, April 10, 2024)

This statement by the reviewer is reasonable in terms of technical accuracy. Nobody *has* actually seen more of the toxic metabolite in the presence of a variety of conditions that cause oxidative stress. But seeing the toxic metabolite might not be enough anyway. Just because the toxic metabolite is produced does not prove that it does any damage. To run the experiment to determine whether the toxic metabolite is produced *and* binds to human tissues, causing damage, babies or small children would need to be exposed to acetaminophen, then samples of their internal organs (called biopsies) would need to be harvested within about two hours of acetaminophen exposure. Samples from healthy children and children with oxidative stress would need to be obtained. Samples of the internal organs of children not exposed to acetaminophen would also be needed as "controls" to ensure that the experiment is valid. In theory, though there would be ethical objections to such invasive procedures to test a drug not shown to be lifesaving, we could get samples from the liver or kidney, but it would be essentially impossible to get samples from the brain, where we think damage is likely. So, in the end, we can probably never get absolute proof that acetaminophen damages the brains of human babies as a result of toxic acetaminophen metabolite production.

But biochemistry, the foundation of the field of pharmacology, tells us that oxidative stress will make acetaminophen more dangerous. The professional toxicologist and the professional pharmacologist who have worked with me for years on this project agree. This was discussed in some detail in chapter 2 of this book. If the body can't get rid of acetaminophen in a safe manner, then the body is stuck with the dangerous pathway, leading to the toxic metabolite. In addition, studies in laboratory animals, discussed in chapter 6, tell us that acetaminophen affects males more than females, and we saw in chapter 2 that people with autism can't process acetaminophen efficiently using safe pathways. For so many reasons, we don't need to give acetaminophen to babies and then take biopsies of their internal organs to prove what is evident from the biochemistry of the system and supported by other observations. The question is, why did the reviewer from *Pediatrics* push against our conclusions based on something that hasn't been absolutely proven, but which absolutely makes sense based on all available information and should be evident to any professional in the field? The point here is that it is true that some things are unproven. But we know enough that we don't *need* more research on this topic. As pointed out in chapter 2, we knew enough by 2009 to take action to alleviate the problem.

As a prelude to chapter 9, which delves into the emotions of science, I have been asked on more than one occasion why I haven't published my work in higher-profile journals. A brief examination of Appendixes A, C and D provides some answers. The scientific jargon is a bit dense, but I think that at least one problem is obvious. When no amount of evidence will convince someone—and they are seeking to refute rather than understand—then there is little that can be done. This situation is termed *invincible ignorance* by mental health professionals, and we saw an example of it in the introduction. But not all is hopeless. We are grateful to the Sapienza University of Rome's Professor Luca Steardo, an editor at *Life*, for looking at the scientific evidence rather than prevailing medical dogma, and publishing our work, some of which was rejected by *Pediatrics*. Similarly, we are grateful to the *Journal of the Academy of Public Health* for publishing our work that was rejected by the *Journal of Xenobiotics*. Further, the flagship journal for the Korean Pediatric Society, unlike its US counterpart, did publish our work.[175] Editors have a lot of power, and their decisions make a difference.

In the next section, I'll review the reasons we think that acetaminophen is responsible for most cases of regressive autism. Regression into autism, resulting in "regressive autism," describes the situation in which a child loses developmental milestones as they literally "become autistic." The fact that regression into autism happens was established by medical science only about three years[176] before Schultz's seminal study in 2008. That study indicated that acetaminophen was the common denominator in children who regressed into autism. However, parents had been reporting regression for decades, and by 1998 that regression had been attributed to vaccination. In this chapter, I will not discuss the role of vaccination in the induction of autism. That will be discussed in chapter 8.

It is easy to see why a parent would associate their child's autism with vaccination. As a scientist, I am not a tremendous fan of studying individual cases. Although such cases can be informative, they can also be misleading, and for that reason scientists prefer to study groups of cases collected in an unbiased fashion. However, when considering *why* a person would think a certain way, I think that specific cases can be useful. Here I will present two cases, both involving parents who watched their child regress into autism. A third case, involving a family with multiple children having infantile autism, is presented in chapter 11.

Case #1

I am thankful to the mother and father of a child who regressed into autism at seventeen months of age in 2001 for allowing me to provide their child's experience as an example. The parents discussed their child's case with me in detail, allowing me to examine their own notes on their child's case as well as the notes from the clinician's office. Although I do not know the father well, I am well acquainted with the mother. She is one of the most conscientious individuals I have ever met and is extremely passionate and highly motivated when it comes to her children's health.

The parents documented ten times in which the clinic office told them their child's reactions to vaccinations or to infections were "typical" or "normal," and that they should administer acetaminophen. The reason the parents took the time to document the clinician's recommendations to use acetaminophen was not because they suspected that acetaminophen was toxic, but rather because it was taken as a sign that the clinic's office believed everything was fine and acetaminophen would be adequate to deal with the symptoms. As we have discussed in some detail,[177] the very idea that acetaminophen might trigger autism did not even exist until the Schultz study of 2008. The parents did not know what we now know. They could not have known, because nobody even proposed the idea until 2008. And, as you will see in this book, there are many reasons why people even today are still unaware of the connection between acetaminophen and autism.

The old phrase "take two aspirin and call me in the morning" was a doctor's way of telling a patient that everything was fine and they can take care of the problem themselves. At least for these parents, the comparable phrase in 2001 was "Everything is normal. Give Tylenol."

The parents were instructed to give acetaminophen to their child every four to five hours, which they faithfully did, literally around the clock, taking pauses only if the baby was sleeping. This treatment regimen usually resulted in administration of Tylenol five times per day, the maximum number of doses currently accepted as the standard of care. The parents administered acetaminophen for teething, infections, and fevers of 99°F and higher. After about six months of age and after a particularly concerning upper respiratory tract infection, they began to use acetaminophen for additional problems, including fussiness and difficulty with sleeping. Unfortunately, the child was diagnosed with chronic ear infections, chronic upper respiratory tract infections, and typically had adverse reactions to vaccines, resulting in fevers, uncontrollable screaming, and reduced sleep that sometimes lasted longer than a week. Based on the detailed records I was provided regarding the child's health and the parents' exact adherence to the physician's instructions regarding the use of acetaminophen, the child was probably given more than a thousand doses of acetaminophen before he showed classic signs of regression into autism at seventeen months of age following acetaminophen treatment for an adverse reaction to an MMR vaccination.

Despite receiving probably more than a thousand doses of acetaminophen, recommendations to take acetaminophen were documented only six times in the pediatrician's records. A typical example is shown in Figure 5.3. The discrepancy between the *actual* amount of acetaminophen administered and the amount documented in the medical record might not be unusual. Such discrepancies obviously create difficulties when assessing associations between acetaminophen use and autism based on data from medical records.

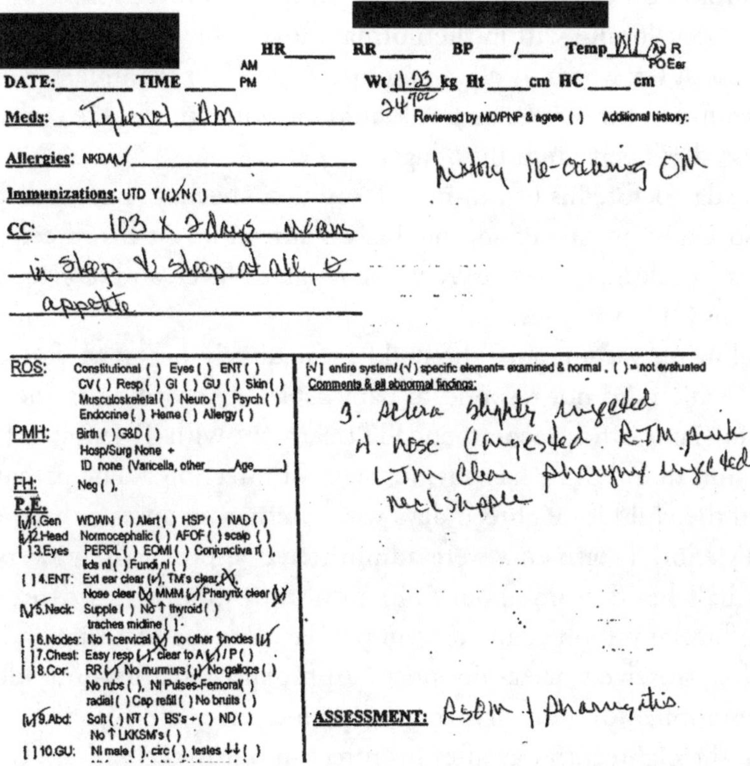

Figure 5.3: Clinic notes from 2001 recording a prescription for acetaminophen (Brand Name Tylenol) for a fifteen-month-old child. The nurse's initials and the date have been obscured to ensure privacy. The child's temperature was recorded at 101.1°F during the visit. Included in the notes are the mother's report that the child was not sleeping well, not eating, and had a fever of 103° for two days. During the visit, the child was diagnosed with a severe infection of the right ear (RSOM = right severe otitis media). The clinician also notes that the child had a history of recurrent ear infections (OM = otitis media). Inflammation of the upper respiratory tract (pharyngitis and nasal congestion) were also noted. The child regressed into autism at seventeen months of age following a severe reaction to an MMR vaccine that was treated with acetaminophen, as noted by the parents and later documented by a specialist.

Case #2

I am grateful to another parent who described his child's regression into autism. The parent provided a very detailed narrative of his child's history, a history that he had revisited extensively over the past twenty-five years. His child's regression occurred in 1999 when the child was fifteen months of age. Again, I am well acquainted with the parent, who is extremely conscientious and devoted to their child. In this case, the parents used baby aspirin from birth through fifteen months of age, including for circumcision, for pain, and for fevers. One of the parents trusted aspirin because that was the drug that they had seen used when they were children, and they had "a bad feeling" about acetaminophen. The child had chronic ear infections from six months until fifteen months of age, but was developing within the normal range. He was crawling and was using about ten words by eleven months of age. For example, when asked what sound a cow makes, he would respond with "moo." He was walking unassisted at twelve months of age.

The day before his fifteen-month wellness checkup, he was diagnosed by a specialist as having an ongoing ear infection. At his fifteen-month wellness checkup, he was fussy, had a low-grade fever, and received DTaP, Hib, and OPV vaccines.

Following vaccination, the child immediately began to run a "high fever," which did not respond to baby aspirin. According to the parent, the clinician's office recommended "treatment with Tylenol and not to bring him in unless he has obvious signs of infection." The parents then treated the child for eighteen days with alternating doses of baby aspirin and Tylenol. Treatments were administered approximately every three and a half hours from about 5:30 a.m. until midnight, or three doses each of acetaminophen and aspirin per day. For the first time in his life, the child received acetaminophen. Approximately fifty-four doses of acetaminophen over an eighteen-day period were administered.

On the eighteenth day after his fifteen-month vaccinations, the child was taken to the pediatrician. The child's fever was resolving at that time, but his bowel movements were no longer solid, and he had lost all words and eye contact. He had also lost "joint attention," which normally develops between nine and twelve months of age. Joint attention is the ability of two people to both focus on a single object of interest, and is a fundamental social and communicative skill. Joint attention is frequently impaired in children with autism. Around that point in time, according to the parent, he also lost his ability to walk unassisted. Less than three months later, the child was diagnosed with autism by his pediatrician, a diagnosis later confirmed by a specialist.

At this point, considering the work of Schultz, who found that the common denominator in regression to autism was acetaminophen, we consider the parents' observations. Surveys show that up to 50 percent of parents who have a child with ASD believe that vaccination contributed to their child's autism.[178] This belief has been widely attributed to a single publication connecting vaccination with regression into autism,[179] but research papers are unlikely to sway public opinion. Indeed, the paper was retracted, and numerous papers have been published claiming that vaccinations do not cause autism. Further, the reputation of the lead author of the paper, Andrew Wakefield, was viciously attacked. It seems unlikely that any research paper, much less a widely criticized paper, could be so influential. A more rational explanation for the persistence of views connecting vaccination with autism is that the real culprit, acetaminophen, hides in plain sight every time an adverse reaction to a vaccine is experienced. The fact that up to 50 percent of parents attribute their child's autism to vaccination is compelling. That number could account for most if not all cases of regressive autism. Indeed, the Schultz study found that the association was almost absolute, with a twenty-fold greater risk of regressive autism associated with acetaminophen exposure. Thus, we deduce that available evidence points toward acetaminophen as essential in the induction of regressive autism.

Again, the evidence from parents and from the Schultz study does not allow us to confidently conclude that acetaminophen triggers *most* cases of regressive autism. Two or three lines of evidence do not make the case. But all of the evidence together, all thirty lines of it described in this book, forms a picture that is clear enough. We conclude that acetaminophen is a trigger for autism at the time of labor and delivery up until about age five, and probably to an extent during pregnancy. Acetaminophen can trigger autism over a wide range of time, accounting for the similarities between symptoms of regressive autism and infantile autism. The only major difference between the two "types of autism," infantile and regressive, is literally the time of induction. The drug essential for the induction is the same, and the resulting mental function is similar.

In the next chapter, I examine scientific work using laboratory animals, which show that acetaminophen would never be approved using today's regulatory standards even for experimental studies in human babies. In other words, if a new drug appeared today that had the same effects on laboratory animals that we see with acetaminophen, the US Food and Drug Administration (FDA) would not let a drug company even *test* the drug on babies, much less approve it for general use. The drug would be dead in the pipeline and never reach market. Here again we will see that breaches of ethical standards of science have affected our understanding of the connection between acetaminophen and autism, but sufficient data has been published so that the adverse effects of acetaminophen on laboratory animals, even at doses lower than those used in human children, are evident.

CHAPTER 6

LABORATORY ANIMAL STUDIES AND EVIDENT TOXICITY

If it's so horrible for rats and mice, why do we give it to our children? You may recall from earlier chapters that acetanilide, a drug the human body converts into acetaminophen, was introduced for use in humans in 1886. One hundred and twenty-eight years later, in 2014, the first study testing the effects of acetaminophen on brain development was published.

Henrik Viberg, an associate professor in the Department of Environmental Toxicology at Uppsala University in Sweden, conducted the experiments on laboratory mice. When doing these sorts of experiments, scientists always make sure to "standardize" the dose of drug so that we can compare exposures between animals and humans. This is done by giving amounts based on body weight. So, if we consider a very young mouse, called a pup, that weighs three grams, and we consider a baby that weighs six kilograms (6,000 grams or about thirteen pounds), we can make everything equitable by giving the drug in amounts proportional to weight. In this case, to give the same dose of acetaminophen to the mouse pup that we give to the baby, the amount given to a baby would be divided by 2,000, because the baby weights 2,000 times more than the mouse pup. An example would be 15 milligrams of acetaminophen per kilogram for a mouse pup, and 15 milligrams of acetaminophen per kilogram for a baby. In both cases, the concentration of drug is the same, and we consider the doses to be the same, even though the baby got 2,000 times more than the mouse pup. From now on, I'll talk about doses in *milligrams per kilogram*, which is abbreviated as *mg/kg*. This will make everything comparable between mice, rats, babies, and adult humans. A normal dose for a baby and for an adult human is 15 mg/kg. However, some babies can get 40 mg/kg if they receive a suppository. When the baby gets a suppository, the actual amount that goes into their system can be highly variable.

Viberg's team exposed laboratory mice to 30 mg/kg of acetaminophen. The mice were ten days old at the time of exposure, and Viberg's group gave the mice two doses of drug, four hours apart. By comparison, most human babies and children get about half that much, 15 mg/kg, but babies can get up to five doses per day, which adds up to more than the mice got. In addition, as I just mentioned, some babies get 40 mg/kg. So, the amount of acetaminophen that Viberg gave his mice, when adjusted for weight, isn't much different than what we give to babies. The results (Figure 6.1) were astounding.

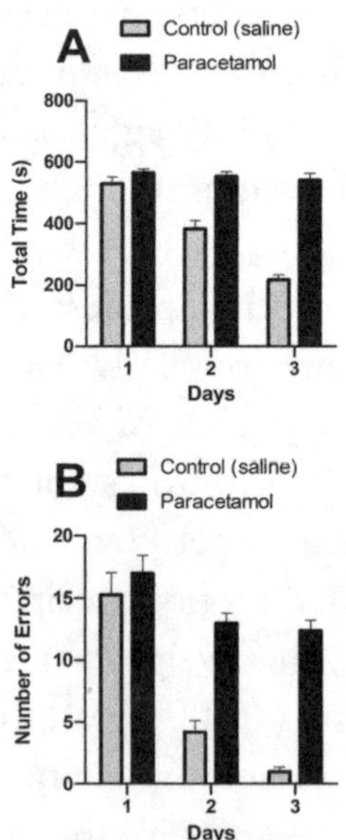

Figure 6.1: Results from the experiments by Henrick Viberg's group in 2014.[180] Healthy mice were treated with acetaminophen (paracetamol) when they were 10 days old and then tested for their ability to find food in a maze when they were juveniles at two months of age. Mice grow much faster than humans and are sexually mature by about two and a half months of age, so two months of age is considered a good time for testing. After two days of training, mice injected with a control liquid (saline) but not mice injected with acetaminophen learned to run the maze faster (A) and with far fewer errors (B). This graph was first published in a European journal,[181] so the European term *paracetamol* rather than the American term *acetaminophen* was used. Both words indicate the same drug. The small lines on top of each bar indicate the degree of certainty or confidence we have in the measurement. We are 95 percent sure that the actual (true) measurement is within the length of that bar in the upward or downward direction.

What you can see in Viberg's study is that mice exposed to acetaminophen very early in life can't think very well at all when they become juveniles. Their brains are rattled. They can't remember a simple maze, and they keep making mistakes. What I find fascinating is that it takes normal mice only two days to essentially learn to avoid almost all mistakes in the maze. That's shown in Figure 6.1B. But if they have been exposed to acetaminophen (called paracetamol in Europe), they can't learn to efficiently run the maze after two days of practice. With that result, modern safety standards dictate that acetaminophen *never* be used in humans during early development, even in an experimental setting. If it does that to a mouse, why give it to a human?

Just before my laboratory at Duke University was shut down, my research group finished running our first set of experiments in laboratory rats using about 12 mg/kg acetaminophen, a dose lower than the dose we typically give to children. You may recall an example I gave of acetaminophen dosing for a human child outlined in chapter 5. The parents stayed up around the clock, giving acetaminophen every four to five hours, up to the maximum number of doses per day, which is five. I followed essentially the same protocol, getting up every five hours around the clock to give the laboratory rat pups acetaminophen. They got five doses per day. In our experiments, we were interested in the effects of acetaminophen on social function. Since rats are much more social than mice, we used rats for our experiments rather than mice. After the rats were exposed to acetaminophen, we let the rats grow up to the juvenile stage and then evaluated their behavior when they encountered a new social partner, a rat that they had never met. Figure 6.2 shows the results.

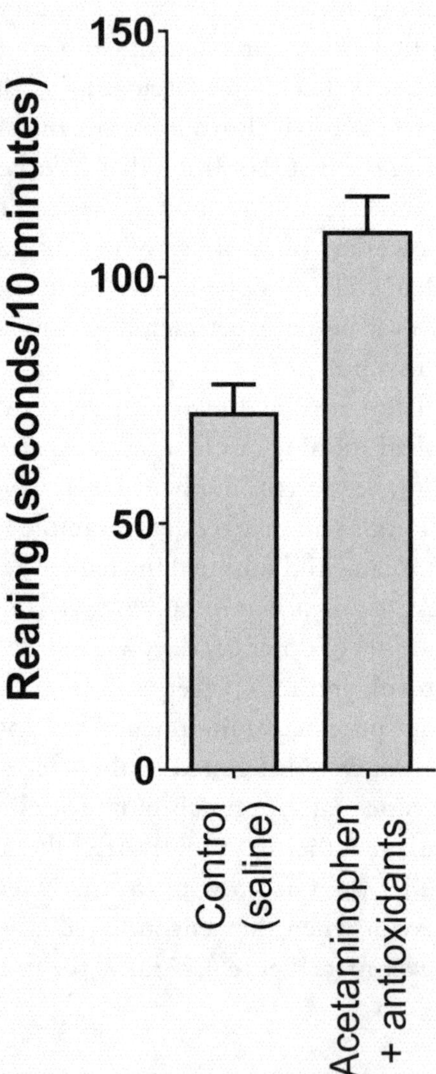

Figure 6.2: Some of the results from our experiments at Duke. Healthy rats were treated with acetaminophen at doses lower than doses used in human babies or were treated with a control liquid (saline). The rats were treated daily from 4 to 10 days of age, and then tested when they were between 37 and 49 days old.[182] In these experiments, acetaminophen was given with antioxidants in a way that should have reduced the adverse effects of oxidative stress. Despite the presence of antioxidants, rats treated with acetaminophen showed more behavior (rearing) consistent with anxiety when meeting a new social partner compared to rats exposed to saline. The small lines on top of each bar indicate the degree of certainty or confidence we have in the measurement. We are 95 percent sure that the actual (true) measurement is within the length of that bar in the upward or downward direction. The chances that the difference is due to a randomly occurring difference in the two groups of rats is about 2 in 10,000 (p = 0.0002).

The rats at Duke exposed to acetaminophen showed a dramatic increase in anxiety-related behavior when meeting a new social partner when compared to their siblings who were not exposed to acetaminophen.[183] This was observed despite the fact that they were given cysteine, an amino acid that serves as an antidote for acetaminophen toxicity.

Some people are surprised to learn that research groups from all over the world, including several laboratories in the United States, have tested the effects of acetaminophen on brain development in laboratory animals. In addition to my laboratory and the Viberg lab, several other examples include the following:

- Margaret McCarthy's laboratory at the University of Maryland gave 40 mg/kg acetaminophen daily to laboratory rats when they were 7 to 13 days of age. That amount of acetaminophen caused long-term modifications to brain development and structure as well as decreased social interactions and sensory function.[184]

- Joshua Herrington's group at Georgia Southern University gave laboratory rats 52 mg/kg acetaminophen when they were 7 days of age, and found changes in the rats' avoidance of open fields later in life.[185] Rats instinctively tend to avoid open fields because an open field is quite dangerous for a rat. In the middle of an open field, rats don't have a place to hide and can be easily caught by a predator. Herrington's observations therefore suggest that acetaminophen exposure early in life impedes development of naturally occurring protective responses in rats.

- Christopher Harshaw's lab at the University of New Orleans gave laboratory mice 104 mg/kg acetaminophen each day when they were 5, 8, and 11 days old. They also included a simulated infection in their experiments, attempting to mimic the situation that often occurs in human babies and children. Similar to the results I saw in my lab using rats, they observed greater avoidance of unfamiliar social partners with acetaminophen exposure.[186]

- L. I. Al-Allaf's laboratory at the University of Mosul in Iraq gave laboratory rats 60 mg/kg acetaminophen daily from 7 to 14 days of age, and observed evidence of damage to the hippocampus, an important part of the brain.[187] (Not that we are aware of any unimportant parts of the brain.)

- Gastaldello Moreira's laboratory at the State University of Londrina in Brazil gave laboratory rats 350 mg/kg acetaminophen daily for approximately 15 consecutive days, starting from 6 days after conception until birth.[188] Moreira's group noted a variety of behavioral changes, including impaired nest seeking behavior, sex-dependent increases in hyperactivity, and less avoidance of open fields.

- Randy Kulesza's team at the Lake Erie College of Osteopathic Medicine in Pennsylvania used the same protocol as Moreira's group in Brazil. They gave laboratory rats 350 mg/kg acetaminophen daily for approximately 15 consecutive days, starting from 6 days after conception until birth. They found impaired auditory function in the offspring.[189]

- Antonio Saad's group at the University of Texas Medical Branch in Galveston, Texas, gave laboratory mice 150 mg/kg acetaminophen daily for approximately 14 consecutive days, from 7 days after conception until birth.[190] They found approximately 30 percent greater average incidence of rearing, an anxiety-related behavior in the male mice exposed to acetaminophen,[191] although they needed more mice to see if the effect was statistically significant.

- Brennan Baker and colleagues, working at Columbia University, gave laboratory mice 150 mg/kg acetaminophen daily starting 4 days after conception until birth, and then continued giving the nursing dams (mouse moms) the acetaminophen until the pups were 14 days old.[192] So, the young mice got some acetaminophen when nursing, although the dose was extremely low. They found that acetaminophen induced changes in gene expression patterns in the brain and in some behaviors, including decreased movements in open fields.

- David Møbjerg Kristensen's lab at the University of Copenhagen gave laboratory mice 50 mg/kg acetaminophen daily starting 7 days from conception until birth, and found decreased male sexual function and altered brain architecture in the adult offspring of the animals exposed to acetaminophen.[193]

Probably one of the most remarkable studies performed in laboratory animals using acetaminophen exposure during pregnancy was conducted by a team at the University of Warsaw.[194] The team, including Dagmara Mirowska-Guzel, chairwoman of the Department of Experimental and Clinical Pharmacology at the university, gave their laboratory rats 5 or

15 mg/kg acetaminophen daily for the duration of pregnancy. This dose is significantly lower than many humans receive, because humans take up to five doses per day of up to 15 mg/kg acetaminophen. Nevertheless, the Warsaw team found alterations in social behavior, biochemical differences in the brain, and differences in brain architecture in the offspring. Alterations in social behavior included an approximately 50 percent reduction in social interactions, even with exposure to only 5 mg/kg acetaminophen. Surprisingly, exposure to only 15 mg/kg acetaminophen daily, the equivalent of only *one* typical dose of acetaminophen per day for an infant, resulted in a more than tenfold increase in a type of wrestling behavior later in life. This result is particularly striking. It is difficult to imagine any mother who, if they had a choice, would want their child to wrestle ten times more than normal.

One of the most important findings to come from laboratory animal work is that acetaminophen has a much stronger adverse effect on the developing male brain than the developing female brain. This observation comes from four different laboratories[195] and applies to both rats and mice. This sex bias toward males seen in laboratory animals matches the sex bias we see in humans with autism, which favors males by about four to one. We consider that finding, from four different laboratories, to be one of our approximately thirty lines of evidence connecting acetaminophen and autism.

In addition, exposure of laboratory rats to acetaminophen before birth causes problems with the processing of sound[196] and with smell.[197] Sensory processing differences, including impairment of hearing[198] and smell[199] are common in autism. We consider this another of the approximately thirty lines of evidence connecting acetaminophen with autism.

Another finding in laboratory animals is that acetaminophen adversely affects gut function.[200] Acetaminophen may also affect gut function in humans, and problems with gut function likely play a role in many cases of autism.[201] In addition, metabolism of molecules structurally related to acetaminophen by gut bacteria are excellent markers for autism[202] (Figure 6.3). It seems likely that these molecules "leak out" of the gut of individuals with autism due to altered gut function, and subsequently deplete the body's ability to detoxify acetaminophen using the sulfation pathway.[203] In addition, an enzyme made by gut bacteria, called BvASST, can remove the sulfate from acetaminophen, undermining the body's attempt to detoxify acetaminophen using the sulfation pathway.[204] The acetaminophen/gut/autism connection, supported by work in laboratory animals, is also a clue pointing toward the role of acetaminophen in the induction of autism.

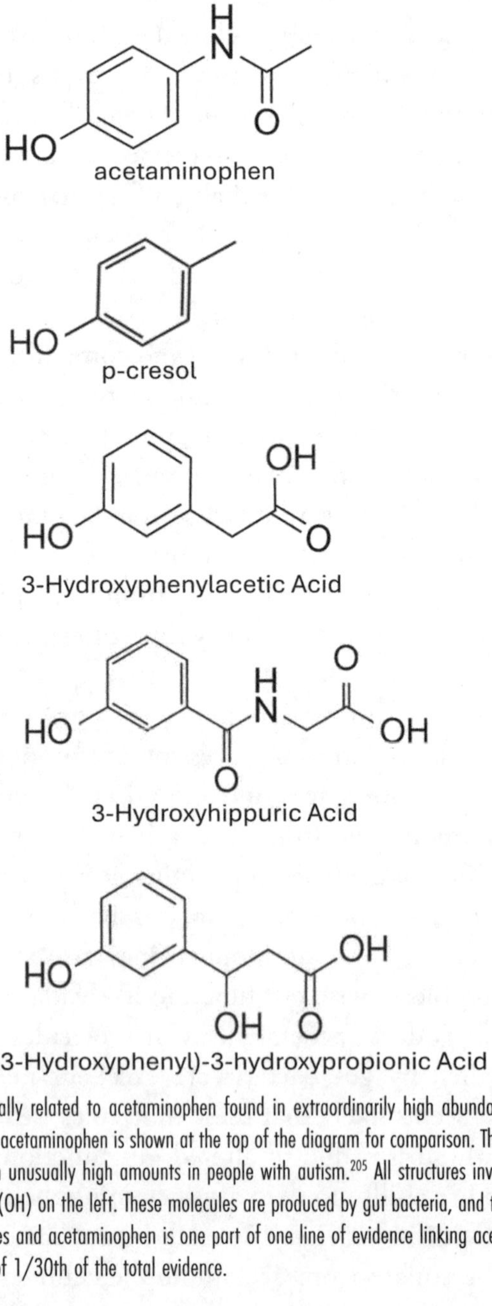
H
N
O
HO
acetaminophen
HO
p-cresol
OH
HO
O
3-Hydroxyphenylacetic Acid
O
H
N
HO
OH
O
3-Hydroxyhippuric Acid
OH
HO
OH O
3-(3-Hydroxyphenyl)-3-hydroxypropionic Acid

Figure 6.3: Molecules structurally related to acetaminophen found in extraordinarily high abundance in individuals with autism. The chemical structure of acetaminophen is shown at the top of the diagram for comparison. The other molecules were previously found to be present in unusually high amounts in people with autism.[205] All structures involve a "phenol group," the ring with the hydroxyl group (OH) on the left. These molecules are produced by gut bacteria, and the similarities between these autism-associated molecules and acetaminophen is one part of one line of evidence linking acetaminophen to autism. In other words, this is one part of 1/30th of the total evidence.

When sorting through the numerous laboratory animal studies evaluating the effects of acetaminophen on brain development, it's apparent that many different laboratories have conducted many different experiments. Experiments have shown that acetaminophen exposure early in life has a wide range of profound, long-term effects on behavior, depending on the experimental protocol. You've seen the very first results ever published, from Viberg's lab in Sweden, and you've seen results from my laboratory at Duke University. *They all indicate that acetaminophen is not safe for the developing brain of mice or rats.*

For the most part, different labs use different procedures, and examine different outcomes, so we don't expect different labs to get the same result. It would be nice to make sure that all results were reproducible. Indeed, reproducibility is considered a cornerstone of science. Especially in the fields of biochemistry and biophysics, the requirement for reproducibility is absolute. Any results published in those fields of science are expected to be reproducible in any laboratory. But sometimes, if the research has implications for public health, experiments can be incredibly helpful even if they are not reproducible. In theory, if we could control every single variable, down to the tiniest detail, then everything would be reproducible. But with laboratory animals, that's just not possible. Neither is it possible with humans.

Even if all our animals were genetic clones, and even if we tried very hard to make the facilities exactly the same, some things would still be different. For example, one technician will smell different than another, and smells affect laboratory animals. So, when testing drug safety with laboratory animals, the goal is not to nail everything down so that a particular outcome is reproducible. The idea is that, before we give it to a single human, we should try giving it to laboratory animals first. If the drug kills the animals or causes severe problems, even some of the time, then we know to avoid exposing humans to it.

After I retired from Duke, I began collaborating with a wonderful team of scientists at the University of North Carolina in Chapel Hill. We conducted a number of new experiments to determine what effects acetaminophen has on laboratory animals. We decided to use somewhat higher doses of acetaminophen than I had used in the past, since adult rats are much more resistant to acetaminophen toxicity than humans are. We picked two doses of acetaminophen, 60 mg/kg and 150 mg/kg.

As we were seeing a number of fascinating and unexpected results, we wanted to know if those results were reproducible. We found that some unexpected and interesting effects were reproducible—and some were not.[206] Among those effects are:

- Acetaminophen exposure early in life affects nursing behavior in laboratory rats. This observation mirrors results of alcohol exposure in laboratory animals.

- Acetaminophen exposure early in life causes laboratory rats to become less responsive to acetaminophen. Similar processes occur in mice and are thought to occur in humans.

- Acetaminophen exposure early in life causes long-term changes that are complex. Sometimes the animals are hyperactive when they grow up, playing more than normal, and sometimes they are hypoactive, playing less than normal.

- The effects of acetaminophen exposure early in life are not always the same and depend on variables that we don't fully understand. One group of rats can be affected one way, and another group of rats can be affected a completely different way.

The study is available for anyone to read,[207] and it has been peer-reviewed, with all reviewers approving of the paper so far as we know. But it was not published in the peer-reviewed literature by the first journal whose reviewers approved of the paper. The story behind this highly irregular situation is a fascinating one and is unique in my experience.

After about four years of work on the project, we decided it was time to write a paper describing our results. We did indeed have some very interesting results and were not convinced that we would learn more by trying to repeat the same studies again. We decided to send the paper to a journal called *PLOS One*, which is a reputable journal run by a nonprofit organization. *PLOS One* had published my first paper showing adverse effects of acetaminophen on brain development in laboratory animals,[208] and they had published another paper on the topic from Antonio Saad's group at the University of Texas.[209] With that in mind, the journal seemed like a reasonable choice. We routinely check journals in advance to make sure we are not wasting our time and the time of the journal staff. If a paper is not a good fit for the journal, we don't bother to try it.

We submitted the manuscript to *PLOS One* on May 8, 2025, and the editorial office assigned it the manuscript identification number of PONE-D-25-24954. According to the journal's website, the journal first conducts a quality control check for "competing interest, compliance with ethical standards for studies involving human and animal subjects, financial disclosures, data availability, and other scientific and policy requirements. Submissions may be returned to authors for changes or clarifications at this stage." On May 14, 2025, we were asked for additional details regarding death of some of the animals as a result of acetaminophen exposure. Some of the animals unexpectedly stopped nursing, which caused death in some cases. Some animals never stopped nursing, while others stopped nursing but regained their nursing behavior and were fine. None of these effects were expected based on anything reported in the scientific literature. We provided the requested information to the journal office the same day, on May 14.

According to the journal's website, "After passing quality control, each manuscript is placed with a member of the Editorial Board who conducts peer review and makes the decision to accept, invite revision, or reject the submitted manuscript." Following procedure, the journal assigned our manuscript to an academic editor, in this case Ali Nemat, an associate professor at the College of Pharmacy at King Saud University. On August 9, 2025, we received two very favorable reviews, and Nemat invited us to revise and resubmit the manuscript. So far, everything was running according to routine. Nothing was unusual. We revised the manuscript and sent it back to Nemat on August 20, 2025. On September 17, 2025, we were informed that one reviewer approved of the revised manuscript, but the other reviewer still had one question that we had not addressed. Nemat again invited us to revise and resubmit the manuscript. We again revised the manuscript, addressing the remaining question, and returned it to Nemat on October 6, 2025. So far, everything was still proceeding as usual.

But world events related to our research changed during the processing of this manuscript. President Trump made his public announcement about the connection between acetaminophen and autism on September 22, 2025, after we had been invited to revise the manuscript a second time (September 17), but before we had returned the revised manuscript (October 6). On November 6, 2025, we were informed by Jianhong

Zhou, staff editor for *PLOS One*, that the manuscript was rejected because it did not meet the ethical standards of the journal. Here is that note.

> In the case of your study, you have not provided us with sufficient details regarding the animal research. Specifically, you have not provided clear scientific and ethical justification for the level of mortality in the study design, and for the large amount of animal death without humane endpoints (lines 227–228). We therefore do not feel your study meets our standards for work involving animals.
>
> We understand that you did seek ethical approval from your local board for these animal experiments. However, we reserve the right to reject any submission that does not meet our internal ethical standards, which in some cases are more stringent than local ethical standards.

This action constituted a deviation from the journal's protocol, which in itself is a type of ethical violation. The review of the *ethics* of our paper had already taken place, we had passed, and we had moved on to the *academic phase* of the review. A journal's established protocols are not meant to be ignored at the whims of journal editors. But did the reason for the rejection actually make any sense? Let us see . . .

I appealed to Zhou, the staff editor, offering to provide more information if needed. I was informed by Jeniffer Corpuz, an editorial assistant, that I could fill out an "appeal request form." I filled out the form and delivered it on November 10, 2025. We had a very straightforward case. Because about 8 percent of the animals in our study died unexpectedly, we were being accused of having poor ethics when it comes to taking care of our animals. However, the animal deaths were unexpected because we were using doses of acetaminophen much lower than were reported to be lethal in the scientific literature. In other words, if we trusted what we saw in the scientific literature about the safety of acetaminophen, then the animals *should not have died*. The animals appeared to be fine in the evening but were dead when we came to check on them in the morning. After we saw this, we adjusted our protocol to alleviate the problem. As I pointed out to the editor on their appeal request form, the pup mortality in the last half of our study was literally better that the current *human* infant mortality in the United States. I understand that this is a sad fact and does not reflect well on the US health-care system, but it is a fact that has a huge bearing on the ethics of our work. Here is what I wrote on the appeal request form, dated November 10, 2025.

> We do understand the concern about the animal death, with 37 out of the 481 animals in our study (<8%) having died without reaching a humane endpoint. But we hasten to point out in the manuscript that the deaths in the first experiments, specifically experiments 1 and 3, were very unexpected. We used a dose of drug (150 mg/kg) that is much lower than the reported LD50 for neonatal rats (1050 mg/kg; Green et al., *Toxicology and Applied Pharmacology* 74, 116–124 (1984)). Despite remarkable results that appear to be clinically relevant (Figure 4A), we switched our approach to minimize death (lines 161–163), reducing exposure of the animals to the drug for experiment 4. This reduced mortality significantly, and in the last series of experiments (experiments 7 through 10), only one animal out of 195 died without reaching a humane endpoint (lines 229–230). This rate of death, 0.51%, is less than the reported infant mortality in the US (0.56%: www.cdc.gov/maternal-infant-health/infant-mortality/index.html).   With this in mind, we believe that our ethical standards, as assessed by the University of North Carolina IACUC (the university ethics committee that oversees and approves all animal work), were adequate. Thus we hope that the editor will accept the paper for publication based on the scientific review, which has been completed after two revisions and three rounds of reviews.

The appeal was rejected on November 17. Here is that rejection.

> Concerns have been raised by the editorial team on the contents of the manuscript relating to the scientific and ethical justification for the level of mortality in the study design. As mentioned in the decision letter, we note the occurrence of numerous animal deaths without the use of humane endpoints. We understand that you did seek ethical approval from your local board for these animal experiments and took steps to reduce mortality, but we do not feel the monitoring described was appropriate. We reserve the right to reject any submission that does not meet our internal ethical standards.

Considering those concerns, the manuscript does not currently meet our criteria for publication requiring that the research meets all applicable standards for the ethics of experimentation and research integrity. This decision has been made on the basis of compliance with journal policy and is independent of any reviewer reports obtained.

The concern is not that the animals did not survive the experiment. What the anonymous editor is arguing here is that we should have *monitored the animals through the night* and then killed the animals quickly and humanely using prescribed methods when they began to show signs of distress indicating they would not survive the study. In their words, they "do not *feel* the monitoring described was appropriate." According to them, it's a *feeling*, not a standard. In my decades of experience, the standard is that scientists often stay up all night for nonhuman primates that are potentially at risk, but not for laboratory rodents that are potentially at risk. Further, I can't imagine that a federal grant would pay the exorbitant costs for personnel to monitor laboratory rodents all night. The right thing to do is to adjust the protocol so that fewer animals die, which is what we did.

According to the journal's published policies, the journal is supposed to be checking for ethical *standards*, not ethical *feelings*.

Adjusting our protocol to obtain animal survival that is better than human infant survival was not enough. In a nutshell, because the editors deviated from their own editorial policy, and because of the way the editors "feel" about monitoring animals, our work cannot be published in their journal, even though the university ethics board approved of the protocol using US federal ethics guidelines and even after the work passed peer review.

This *PLOS One* editorial decision means that dissemination of new findings regarding the impact of the world's most popular drug on brain development will be delayed. The review process wasted more than six months of time just when public awareness of the issue had reached an all-time high. I understand how this situation could enrage anyone who has a child with autism and is passionate about research on its causes. With one child being diagnosed with autism in the United States about every five minutes, six months is a lot of time to waste.

At the same time, I appreciate concerns related to laboratory animal welfare. I worked in the field of transplantation research for decades, where animal welfare is especially challenging. If one stitch or one blood clot shifts the wrong way at the wrong time, an animal can unexpectedly die. We see the same thing with the surgeries in humans. Neither research nor clinical practice is without risk. We do our very best to avoid that risk, but complete avoidance is impossible. The goal is to improve human life, and to respect the animals we use in our studies as we try to achieve that goal.

My main concern regarding the ethics of laboratory animal research is that we have a vast amount of data from numerous laboratories showing that acetaminophen is dangerous for brain development. And none of it has affected public policy. If the results are ignored, then there is absolutely no point in doing the research. In addition, if journals subvert the scientific process and refuse to publish research they don't "feel good" about, then there is no point in doing the research. It is unethical and even immoral to use laboratory animals to do research that is pointless. Even the poisons used to control wild rats in New York City are not employed without purpose. The New York rat population, about two million in number by a 2014 estimate,[210] is controlled in part using poisons that kill the animals without a humane endpoint as defined by laboratory animal research standards. Countless more die of starvation and predation, neither of which is defined as a humane endpoint by laboratory animal research standards. But there is a purpose, whether it be pest and disease control or the natural course of rodent life in the city. Without a purpose, there can be no justification for animal use, regardless of whether we feel good about our ethical standards.

From an animal welfare perspective, the decision by *PLOS One* editors is difficult to rationalize. Based on the available published literature at the time of the editor's final decision, nobody could know that acetaminophen at the doses we used might result in the death of laboratory rat pups. Thus, until our study or a similar study is published, other investigators will not know to avoid experimental protocols similar to ours—potentially leading to *more* unexpected animal deaths. Further, President Trump's announcement could stimulate more research on the topic, resulting in more research and thus even more unexpected animal deaths. After the president's announcement, it was more important than ever that these new findings be published, if the safety of laboratory animals were indeed the primary concern.

While the editorial decision was unethical and could conceivably cause harm if it delays public awareness of research on the connection between acetaminophen and autism, it was not necessarily irrational or immoral. Although the timing of the rejection after peer review was unreasonable and the excuse for the rejection is, at best, difficult to accept, rejection of the manuscript for any reason was likely justified in the minds of the editors. As I pointed out in the introduction, subversion of the scientific process is often perpetrated by individuals who think they are doing the right thing from a moral perspective. I believe that the editors did not want to be involved in research they believed could contribute to public harm. If they believed that acetaminophen is safe for babies, despite the obvious harm it does to laboratory animals, and altered the scientific review process in response, then their dogma trumped their science. Again, as pointed out in the introduction, I am not saying the editors were free from conflicts of interest, emotional compromise, or cognitive dissonance. I suspect they were swimming in all of that, an issue that will be covered in chapters 9 and 10.

In the next chapter, I will cover some extremely compelling evidence connecting acetaminophen and autism that we have not yet covered. Some of the evidence is shocking even to most pediatricians.

CHAPTER 7

MORE EVIDENCE, FROM THE AMAZING TO THE OBSCURE

Every veterinarian knows that cats cannot tolerate acetaminophen. The drug is toxic to adult cats. In my experience, most cat owners are also quite well aware of this fact. What fewer people know is the *reason* why cats cannot tolerate acetaminophen. They are deficient in a metabolic pathway that safely eliminates acetaminophen from the body.[211] This pathway, the glucuronidation pathway, was considered in chapter 2. Newborn babies are deficient in the same pathway.[212]

With that evidence alone, the obvious question is, why do we give babies acetaminophen? If we know that cats have a problem with acetaminophen that makes it toxic, and we know that babies have the same problem, what are we thinking?

We do, in fact, know the answer to that question.

The answer is, quite simply, that we, as a society, *believe* acetaminophen is safe. It is part of our collective consciousness.

If you really believe that acetaminophen is safe for babies, and if you know the information about cats and babies having the same deficiencies that make acetaminophen toxic to cats, how do you reconcile this information?

One way to reconcile the situation is to do what one anonymous "expert" reviewer for the journal *Pediatrics* did (Appendix D). He or she simply created an *imaginary* metabolic process in babies that compensates for their lack of glucuronidation. The events surrounding our article submitted to *Pediatrics* were described in detail in chapter 5. Here, from

Appendix D, is how our expert reviewer from *Pediatrics* describes their imaginary metabolic process that protects babies from acetaminophen:

> . . . in fact, children in general have a higher capacity to sulfate and therefore are generally relatively protected from APAP (acetaminophen) toxicity . . .

In response to the reviewer's imaginary safety net for babies, our research team responded with the following facts, including references.

- The reviewer is apparently confused by the fact that babies, with glucuronidation capacity being poor, have *only* sulfation left as the major, safe pathway to get rid of acetaminophen. This does not mean that sulfation is *enhanced* in babies. It means that it's all that is left.
- Analysis of acetaminophen metabolism in babies shows that it is, in fact, impaired compared to metabolism in adults.[213] The reviewer's safety net for babies is imaginary.
- Analysis of acetaminophen metabolism in babies shows that the drug does, in fact, get shunted down the toxic pathway more in babies[214] than in adults.[215] This is more evidence that the reviewer's safety net for babies is imaginary.
- Sulfation in children with autism is known to be impaired based on multiple, independent studies. This fact was recently reviewed by Richard Williams.[216]

The expert reviewer for the most influential pediatric journal in the world has faith in an imaginary safety net that can be described as "enhanced sulfation." Babies have some sulfation, but it's not enhanced. More importantly, children with autism tend to have impaired sulfation, as described in chapter 2. So, not only does the imaginary safety net not exist, but the pathway that the reviewer believes to be enhanced is actually impaired in many children, making them even more susceptible to injury. Richard Williams, with no university affiliation or evident scientific credentials, has a very understandable review of the topic in the introduction of his paper.[217] (For unknown reasons, Williams does not cite some work on the topic from academic institutions, including the first study from 1999[218] and other important studies[219] showing that children with autism have an impaired sulfation pathway. If I had been a reviewer on Williams's paper, I would have suggested that he add those articles in his introduction.) It is unfortunate and indeed tragic that an expert reviewer who serves as a gatekeeper for the world's most influential journal in the field of pediatrics does not understand metabolism in children with autism as well as Williams, who has no apparent professional training.

The full text of that correspondence with the expert reviewer is in Appendix D. If you examine Appendix D, you may be distracted by the reviewer's repeated complaint that we are citing our own work. It is indeed bad form for a scientist to cite his or her own work if somebody else can be cited. But I'll remind the reader that the editor required us to condense down our manuscript. During the process of condensing our manuscript, the editor allowed us to have only seven citations. As you will see in this chapter, most of the evidence I present is not from my laboratory. For example, I have never worked with cats in the lab or studied their metabolism. But since the editor only allowed us seven citations, we had no choice but to refer to reviews that contain the full litany of relevant references. For whatever reason, my group is the one publishing the most comprehensive reviews. Apparently, the reviewers were unaware that the editor had only allowed us seven citations, so they berated us repeatedly for citing our own reviews. Please don't let that distract you as you glance over Appendix D.

Back to our story: The bottom line here is that when experts start making up scientific-sounding things out of thin air to reconcile their beliefs, we have a problem. More specifically, the progress of science has a problem. My research group saw the same pattern when systematically examining the treatment of seminal work that is difficult to explain unless acetaminophen is, in fact, important in the induction of autism.[220] The study by Schultz, published in 2008 showing a dramatic connection between acetaminophen use in children and regressive autism,[221] is an example of a paper that has been dismissed for scientific-sounding reasons that have no validity.[222] As another example, the Danish study showing a connection between circumcision and autism,[223] along with its implications for the connection between acetaminophen and autism, are routinely dismissed with scientific sounding but senseless logic.[224] One group in particular, led by Brian Morris at the University of Sydney, levied approximately twenty relatively distinct criticisms[225] against the Danish study in various combinations of up to ten distinct criticisms at a time in six different publications.[226] But none of the criticisms had any validity.[227] In defense of Morris, he apparently has no qualms with the view that acetaminophen triggers autism, but he seems to be very opposed to the idea that circumcision itself triggers autism. He is almost certainly right about that.

And, of course, the conclusions we draw are supported by the Danish study regarding circumcision, but depend on a large body of supporting evidence, not on any one study. I hope the reader will forgive me for repeated reminders of this fact that is so obvious to many. But scientists reading this book need to be constantly reminded. Our scientific training has programmed us to dig deep into a single study and focus on its weaknesses. We are trained to look at the small picture, even when it causes us to miss the big picture. The training is intense, and the programming is effective, so we need frequent reminders that the approach we were taught, often called *reductionism*, sometimes isn't helpful for solving real-world problems. For those of you who do not share that programming, please forgive me for repeating this mantra so many times.

The next question is, why do experts believe that acetaminophen is safe for babies so strongly that they need to make up scientific-sounding nonsense to justify their beliefs? Where did that belief come from? We know the answer to that question. The answer is, because the peer-reviewed medical literature says it's safe. That's why. Science is the foundation of medicine, and since the peer-reviewed literature, which is based on science, says acetaminophen is safe, then it must be safe. That was the belief. But why did the peer-reviewed literature say acetaminophen was safe?

A brilliant member of my team at the time, who wishes to remain anonymous, suggested that we needed to prove that acetaminophen was never proven safe. Proving a negative is very tricky. The only way to perform such a feat is to systematically turn over every single study ever done to see if anybody ever checked to see if acetaminophen is safe for brain development. I worked with a brilliant professor of information science, Vincent Larivière at the Université de Montréal, to conduct the research.

The technique we used is called *citation tracking*. Basically, we start out with a *systematic review*, which means that we use a systematic method to search the literature for a specific topic. It's the same as searching a field for a lost coin by marking off every square foot of the field, and then checking every square foot, one by one. In this case, we used automated tools to search an electronic database for anything having to do with the safety of acetaminophen for babies or children. That yielded thousands of papers, which my research team and I methodically screened. That screening yielded hundreds of papers that said acetaminophen is safe for babies and children when used as directed. The final step was to look at those hundreds of papers and track down what support or proof they had for their statements that acetaminophen is safe for babies or children. The method is extremely time-consuming but is simple in conception. Figure 7.1 shows how it is done.

STEP 1

Electronically search the medical database for every paper that mentions anything about safety of acetaminophen in babies or children

STEP 2

Manually screen each of those thousands of papers to find the ones claiming that acetaminophen is safe for babies or children.

STEP 3

Manually screen each of those hundreds of of papers to see what evidence they provide for their statement that acetaminophen is safe for babies or children.

Figure 7.1: The method of systematic review with citation tracking.

Steps 1 and 2 shown in Figure 7.1 comprise the "systematic review" part of the work. This part of the process allows us to find most of the papers ever published that are in the database and claim that acetaminophen is safe for babies and children when used as directed. Step 3, the most time-consuming part, is the "citation tracking" part of the project. That part of the project can expand dramatically. One paper might cite another paper, which might cite another paper, and so on. Figure 7.2 shows you how tracking down just one paper can expand dramatically.

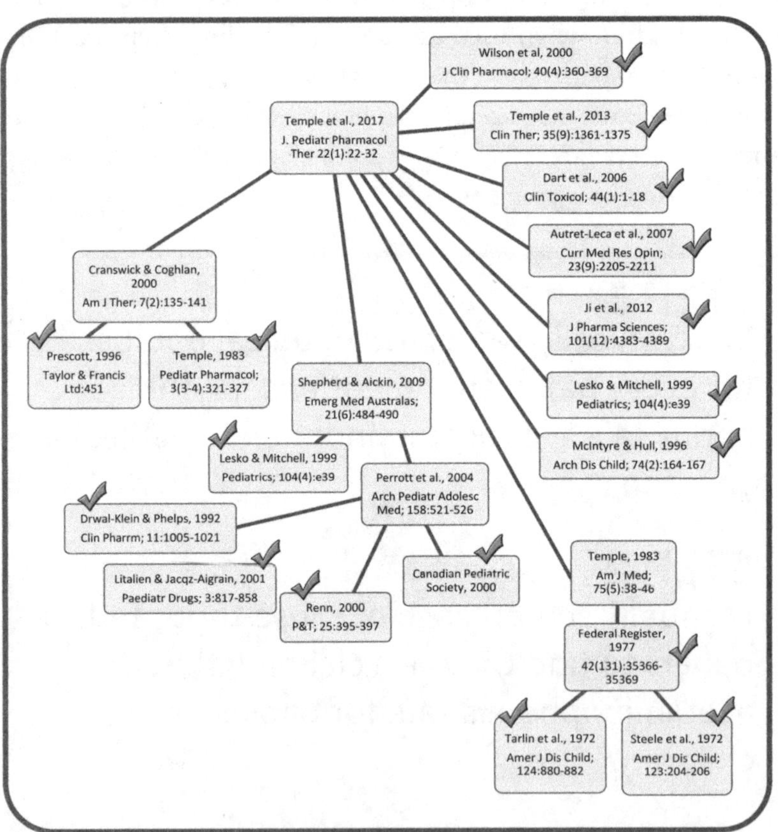

Figure 7.2: Citation tracking from one article, Temple et al., 2017. Articles describing new experiments designed to test safety of acetaminophen or which contain claims of safety without citation are indicated by a check mark.

After more than eight months of labor to track down every single study and evaluate those studies, we found that only fifty-one experiments were ever conducted looking at the safety of acetaminophen, and the typical time of follow-up was only two days. Nobody had ever checked for brain function later in the life of babies or children exposed to the drug—at least not in a way that indicated that the drug is safe. The assumption was made that babies are just small adults and interact with acetaminophen the same as an adult. We know this is not the case, and even in the 1980s, when many of those safety studies were being done, most people knew it was a bad idea to assume that a baby is a small adult.

Acetaminophen overdose is a leading cause of liver failure in adults and is often fatal.[228] So, because adults show signs of liver damage when they have too much acetaminophen, it was assumed that babies would also show signs of liver damage if they got too much acetaminophen. We now know that this assumption was not valid. A 1984 study would eventually show that, in very young laboratory animals, even lethal doses of acetaminophen do not damage the liver.[229] This false assumption is the first of the great three scientific errors blinding scientists to the connection between acetaminophen and autism. I have already discussed the other two. To review, this time in chronological order:

- The first mistake was to assume that babies are the same as small adults, and to only check for liver function before deciding acetaminophen is safe for them.
- The second mistake was to focus on pregnancy, when risk from brain damage is probably lower than at birth and in early childhood. (That was covered in chapter 5.)
- The third mistake was to treat interacting variables as confounding factors (That was covered in chapter 3.)

As you have seen in this book, a fourth mistake was to repeatedly ignore observations that didn't fit the idea that acetaminophen is safe. Basically, the scientific process could have overcome those three errors, but the scientific process was subverted, thwarting efforts to reveal and rectify errors.

After all that citation tracking, we had proof that there was no proof. We found ourselves looking at the smoking gun telling us exactly *why* people had believed acetaminophen was safe for babies and children in the first place. The most cited paper prior to 1995 used as evidence that acetaminophen was safe for babies and children was a 1978 editorial in the journal *Pediatrics*.[230] That editorial simply noted that pediatricians were not finding much liver damage in babies exposed to acetaminophen.

You may recall from earlier in this chapter that the same journal has been less receptive of current work showing—without reasonable doubt—that acetaminophen is not safe for brain development.

We wrote the manuscript describing the main result, which was that acetaminophen was never proven safe for brain development. After much discussion, we decided to submit the paper to a journal called *Systematic Reviews*. The paper was submitted on November 10, 2020, but was rejected by the editor on April 14, 2021. The reviewers objected in large part because the paper was technically not a systematic review, but rather was more of a citation tracking paper. The reviewers also didn't like the fact that we focused on studies saying that acetaminophen is safe. They wanted us to survey everything, including studies saying that acetaminophen is not safe. This approach would be typical of a systematic review, which often compares evidence for or against a certain viewpoint. Since nothing in the reviewers' comments was technically wrong, there seemed to be no point in arguing with the editor.

We tried another journal. The publisher had suggested we try *BMC Pediatrics*, which is a prestigious journal. We submitted the manuscript to *BMC Pediatrics* on June 1, 2021, and received two reviews from anonymous reviewers on September 3, 2021. One anonymous reviewer stated that "Although the intention was good and the work in undertaking this study was admirable, the conclusion that the drug was never proven to be safe was not totally unexpected." The reviewer went on to add that "In fact, most of the drugs used in children have never been really assessed on all the safety aspects."

The second reviewer for *BMC Pediatrics* was also matter of fact:

> This is an extremely important study stating that paracetamol affects neurodevelopmental outcome in newborns and infants when given during pregnancy or to infants. The manuscript has clearly demonstrated that the mistaken belief that paracetamol is safe to be given in infants is based on studies that have either only looked at liver function or have had too short a follow up period with neurodevelopment not being assessed. However, I find much of the manuscript repetitive and too detailed with concentration on a detailed methodology . . .

The editor assigned to the paper, Daynia Ballot at the University of the Witwatersrand in South Africa, rejected the paper based on those comments. We could not argue. The results *were* expected as pointed out by the first reviewer, and the paper was boring, as pointed out by the second reviewer. My team has written papers that are not boring, and this was not one of them. Looking at pages and pages of tables showing that nobody ever tested a drug for brain development is never going to make for an exciting day. The fact that the results were important was acknowledged by the second reviewer, but we couldn't get around the first two issues. The results were a foregone conclusion for anyone who knows the field of drug testing, and the paper was boring.

We tried another journal. On October 13, 2021, I submitted the paper to the *European Journal of Pediatrics*. The editor assigned to the paper was Peter de Winter, chief editor of the journal and dean of the Spaarne Gasthuis Academy.

On November 2, 2021, we received one anonymous review. The reviewer brought up a number of points, which were all invalid:

- The approach was "bizarre," since the safety assessments were never meant to test for neurodevelopment.
- Regulatory agencies say that acetaminophen is safe. (This point was repeated twice.)
- Laboratory animal studies are not all run the same way.
- Studies of exposure to acetaminophen during pregnancy are inconclusive.
- Some limited data do exist showing that acetaminophen is safe for neurodevelopment. In particular, studies from Outi Aikio's group at the Oulu University Hospital in Finland[231] are "reassuring."

The editor, de Winter, rejected the paper based on those comments. But this was different from our previous reviews. Here we had a review that we could really sink our teeth into. We couldn't argue that the paper wasn't boring, and we couldn't argue that the results should be surprising. But this expert reviewer was saying things we could refute. Going down the list, first we could argue that it was precisely because people *believed* acetaminophen was safe that we needed to do this unusual study. Second, we could argue that trusting the opinions of regulatory agencies was a logical fallacy (argument from authority), and third, multiple laboratory animal studies using different methods all pointing to the same conclusion is a strength, not a weakness. Finally, we knew a great deal about those studies during pregnancy and those studies from Oulu in Finland. They were more concerning than the reviewer let on. For example, the study from Oulu actually showed double the rate of autism with acetaminophen exposure, even though the antidote for acetaminophen poisoning was added to the formula given to the babies. The only way to know that information is to dig into the paper and look at the actual data, unfortunately. The abstract of the paper reveals nothing.

So, with this in mind, I sent a protest letter to de Winter. I sent the letter the same day I received the rejection, on November 2, 2021. I warned my students that protesting an editor's decision almost never works and would likely be a waste of time. But we had to try. We had to give the editor a chance to look at the evidence.

De Winter's response, received the next day, on November 3, 2021, was straightforward: "Thank you for your email. After reading your interesting reply, I grant you the opportunity to rewrite according to the suggestions and questions of the reviewer and resubmit your manuscript."

After going through the review process at the *European Journal of Pediatrics*, the paper was accepted on February 1, 2022, about fourteen months after we had first submitted the paper for publication to a journal. The expert reviewer from *BMC Pediatrics* as well as any expert in the drug testing field already knew that acetaminophen was never shown to be safe for brain development, but Peter de Winter allowed us to put that information into the scientific literature, where it desperately needed to be. Beautifully done, Dr. de Winter. Peter de Winter's decision to look at the evidence rather than to blindly follow an expert who provided only demonstrably false information was refreshing. We now had the tool we needed to confront the field of medicine with the dogmatic nature of their belief in the safety of acetaminophen. That paper [232] is one line of evidence out of approximately thirty.

When I give lectures about the connection between acetaminophen and autism, I must select an order of evidence. I always start out with the metabolism of the drug, showing that children with autism have metabolic problems which force acetaminophen down a toxic pathway. Then, much as I have done in this book, I take a look at associations between acetaminophen and autism through time, and I explain the problem with confusing confounding factors with interacting variables. At this point, when I'm about halfway through the evidence, I add the next line of evidence. When talking to a group of educated people, this next line of evidence usually gets most people past their skepticism into the realm of understanding.

In adults, acetaminophen blunts social awareness. That's a big line of evidence. That line of evidence tells us that, in some way we don't yet understand, acetaminophen affects the same thinking processes in adults that are affected in people with autism. People with autism have impairments in social awareness. That trait is the most well-known feature of autism spectrum disorders.

Several academic clinicians and scientists have examined the effects of acetaminophen on adult cognition. Nathan DeWall, a professor of psychology at the University of Kentucky at Lexington, was the first to examine the effects of acetaminophen on social awareness in adults. His team's first paper, published in 2010, showed that acetaminophen temporarily blunts the pain of social rejection.[233] Ian Roberts, a scientist at the University of Toronto, working with investigators at The Ohio State University, including Baldwin Way, a professor of social psychology, found that acetaminophen may alter social trust in adults.[234] They found that, in surveys, acetaminophen use was associated with decreased social integration in the neighborhood and decreased trust in neighbors. Daniel Randles, a social psychologist at the University of British Columbia, found that acetaminophen blocked normal responses to adverse social stimuli.[235] Baldwin Way, the aforementioned professor of social psychology, also found that acetaminophen blunts emotional responses to both positive and negative stimuli.[236] Way speculates that acetaminophen is a kind of "social analgesic" that somehow doesn't affect our logical thought processes related to social issues but affects our *perception* of social issues.[237]

This bit of evidence has a pattern shared by several other lines of our evidence. Since A is connected to B, and B is connected to C, then A might be connected to C. Since acetaminophen is connected to alterations of social awareness, and autism is connected to alterations of social awareness, then acetaminophen may be connected to autism. The same pattern holds with gut function. Since acetaminophen is connected to gut function,[238] and autism is connected to gut function,[239] then acetaminophen may be connected to autism. In the case of gut function, we can add that molecules shed from the gut with chemical structures related to acetaminophen are associated with autism.[240] Those molecules ("phenols") were described in chapter 6.

This type of evidence is circumstantial. Here I could give my usual disclaimer and point out that numerous other lines of evidence are also important, and that many of those are *not* circumstantial. But I think first we should consider the importance of circumstantial evidence. Let's return to our easily forgotten burglar, Joe Banana, who appeared earlier in this book. Let's say that an eyewitness says that the burglar of the gumball machine was wearing a green hat and was accompanied by a woman about six feet (1.8 meters) tall. If Joe happened to be wearing a green hat when the burglary was committed, and he happened to be in the neighborhood when the burglary was committed, and he happened to be hanging out with his friend Amy Tomato, who just happens to be about six feet tall, then we have circumstantial evidence pointing toward Joe as the criminal.

The criminal investigator sees the same circumstantial pattern in his investigation that we see with alterations in social awareness and with gut function. If somebody with a green hat stole the gumballs, and if Joe Banana was wearing a green hat, then maybe Joe Banana stole the gumballs. It's circumstantial.

Circumstantial evidence adds up. Most law enforcement officers, given all the evidence above, would guess that Joe was probably the gumball thief, especially if green hats weren't being worn by numerous local people for a special occasion such as a sporting event. If the stakes were higher, for example if a child's life was at stake, circumstantial evidence would lead to a rapid and thorough investigation of the suspect, who would likely quickly find himself behind bars, where he would finish out his life. If we take the circumstantial evidence seriously, it can be helpful. But it's not proof.

Another way to look at this situation is that we already have an overwhelming amount of evidence. Additional circumstantial evidence only helps us make sure we are on the right trail.

Now let's look at one bit of evidence that is not quite as circumstantial as the evidence related to acetaminophen-mediated social impairment. While working at Duke, I learned that, even though they have oxidative stress, people with cystic fibrosis are super-metabolizers of acetamino-phen.[241] In other words, scientists had determined that people with cystic fibrosis metabolize acetaminophen more safely and more rapidly than people who do not have cystic fibrosis. This probably has to do with their bodies trying to compensate for all the oxidative stress they have in their lungs and gut. This situation puts people with cystic fibrosis in stark contrast to others who have problems with oxidative stress. In fact, numerous conditions associated with oxidative stress are associated with a high prevalence of autism, as we would expect.[242]

Regardless of the reason why, scientists documented the observation that people with cystic fibrosis were super-metabolizers of acetamino-phen. It also turns out that, because cystic fibrosis was essentially a death sentence twenty years ago, the mental health of people with cystic fibrosis has been carefully monitored. It seemed reasonable that, if people with cystic fibrosis were super-metabolizers of acetaminophen, they should be protected from autism spectrum disorder. Our team could not find any mention of autism in patients with cystic fibrosis, despite the fact that their mental health was monitored carefully. I talked with lung trans-plant surgeons who had worked with cystic fibrosis patients every day for decades, and none remembered *ever* seeing a patient with autism. I talked with people within the cystic fibrosis community, and none knew of anyone with autism.

We are not saying that nobody with cystic fibrosis has or has ever had autism. We are, however, saying that it seems that people with cystic fibrosis have a lower prevalence of autism than average.

This observation of a low prevalence of autism in people with cystic fibrosis, published in 2017,[243] is still circumstantial, but it's relatively strong circumstantial evidence. This is akin to the police finding some gumballs in Joe Banana's pocket that match the gumballs that were stolen. While this is not so easy to explain away, it's still not proof. Maybe Joe Banana bought the gumballs from a local vendor. Similarly, maybe somehow we just couldn't find the people with cystic fibrosis and autism, or maybe some other reason besides acetaminophen metabolism explains their low prevalence of autism. This evidence is not proof, but it's compelling.

It would be nice to have one piece of evidence that was absolutely, positively, perfectly convincing. But even if Joe's fingerprints were all over the gumball machine, and the gumballs in Joe's pocket matched the gumballs taken from the machine, it's still possible that Joe was trying to buy gumballs from the machine, and the gumballs in his pocket just fell out of the machine. We would really need a video of Joe Banana stealing the gumballs if we wanted absolute proof.

When my research team drew the conclusions we drew, we knew that absolute proof was unobtainable. But we also knew that the amount of evidence available is overwhelming, to the point where any reasonable, unbiased person, when they examine the evidence, will draw the same conclusions that we have drawn. Scientists generally don't need absolute proof. We just need enough evidence for or against a hypothesis to draw a conclusion.

Let's leave Joe Banana to discuss his situation with the local police and move on to another wonderful line of evidence. For years, I was asked on a regular basis, "How can a single drug cause such a complex spectrum of symptoms?" The answer I gave is still valid: Brain development is a very complex process, so if we throw a toxic chemical into the complex process, we can get a lot of different and unpredictable problems. It's the same as throwing a bit of sand into a very complex clock. Fifty different things could go wrong. That answer is still a good answer, especially considering that every person is different in terms of their genetics and their environment. Even identical twins aren't exactly the same.

But then somebody handed me some excellent information that helped answer the question much better. They had worked with children with disabilities for years and had a broad perspective on child development. They explained to me that autism spectrum disorder shares many similarities with *fetal alcohol spectrum disorder*. My team published this observation,[244] and it adds to our list of evidence. This evidence does not tell us *how* a single drug induces a complex spectrum of symptoms, but it tells us that a single drug *can* induce a complex spectrum of symptoms. If we want to think about how this happens, we might go back up to the analogy of throwing sand into the complex clock.

Although fetal alcohol spectrum disorder and autism spectrum disorder are very distinct from each other, the list of similarities is remarkable. Here are some of them.

- Both disorders are characterized by a spectrum of conditions, and both disorders can be difficult to diagnose in some cases.[245]
- Both disorders are highly variable in terms of cognitive impairment.
- Both disorders can involve sensory and motor difficulties,[246] attention-deficient/hyperactivity disorder–like symptoms, including executive dysfunction, attentional deficits, and impulsivity,[247] other comorbid mental illnesses such as anxiety, mood disorders, obsessive-compulsive disorder, etc.,[248] and intellectual disability.[249]
- Both disorders can be associated with other medical conditions such as high rates of seizures, sleep problems, abnormal eating behaviors, and disorders related to immune function.[250]
- Both disorders are known to have many risk factors which contribute to susceptibility, including genetic factors[251] and environmental factors, such as those relating to parental age, health, nutrition, and other prenatal and postnatal factors.[252]
- Both disorders are triggered by exposure of susceptible individuals to drugs (acetaminophen and alcohol) that block pain and fevers.[253]
- Both disorders are triggered by exposure of susceptible individuals to drugs (acetaminophen and alcohol) that are metabolized by the human body into molecules (quinoneimines and aldehydes) that belong to a very specific class of toxic substances called "highly reactive electrophilic compounds."[254]
- Both disorders are triggered by exposure of susceptible individuals to drugs (acetaminophen and alcohol) that alter nursing behavior in laboratory animals.[255]
- The commonly recommended treatment for both disorders is early intervention,[256] akin to rehabilitation after any type of brain injury, although the success of the treatment is highly variable in both cases.

The main point here is that a single drug, related to acetaminophen, is already accepted as a cause of a complex spectrum disorder. Again, we are not by any means saying that fetal alcohol spectrum disorder, previously known as *fetal alcohol syndrome*, is the same as autism. Clearly this is not the case. The point is that these two spectrum disorders share many similarities, adding to the body of evidence connecting acetaminophen with autism.

Sometimes I am given a bit of evidence that is consistent with the connection between acetaminophen and autism, but perhaps not extremely compelling. But evidence is evidence. If I found some evidence telling me that acetaminophen is not connected with autism, I would definitely publish that. I would be morally and ethically obligated to publish that evidence, if I ever found any such evidence. So, to be fair, I should publish whatever evidence I have connecting acetaminophen with autism.

Here, in Figure 7.3, is a diagram showing that the amount of acetaminophen sold in two Scandinavian countries was, at one point in time, associated with the amount of autism in those countries. (See also Figure 4.1.)

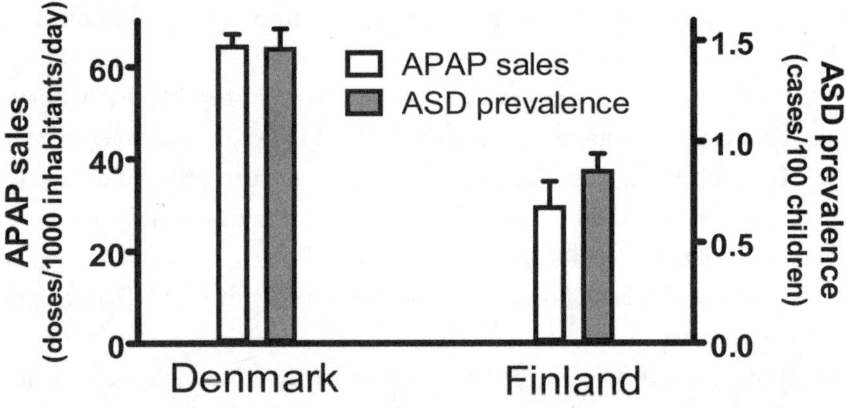

Figure 7.3: The sales of acetaminophen (APAP) in Denmark and Finland correlated with the prevalence of autism spectrum disorder (ASD) in those countries. This graph was published in the journal *Children*, and more details about the data can be found there.[257]

When I publish something like this, I know I am tempting people to give me a lecture, explaining that association does not equal causation, and that this association proves nothing. Freshmen college students and experts on social media with interesting pseudonyms are especially tempted. So, when I published this graph, I made sure to include several obvious limitations to the observation and proclaim, "When considered in light of other lines of evidence, this association (Figure 7.3) adds to the burden of evidence demonstrating without reasonable doubt that acetaminophen use in susceptible babies and children causes many if not most cases of ASD."

I would think that this statement in the paper would have reduced the temptation to give me the lecture about association and causation, but it certainly didn't eliminate it entirely. I still got the lecture.

This association is just one more association among many and doesn't add much to our total weight of evidence. At the same time, I realize that the numbers in this association represent a tremendous number of cases of autism in Denmark, and I do not in any way want to diminish the importance of the suffering that autism can cause families, especially in cases of severely affected children. But from an evidence perspective, this evidence is just another association, and we already have a lot of those.

If you have read this far, you've seen almost all the evidence. I've left out some biochemical and cellular evidence, and, frankly, it's not much different than the rest of the evidence you've seen. For example, studies show that acetaminophen or acetaminophen metabolites produced by the human body damage growing brain cells in tissue culture,[258] which indicates (once again) that acetaminophen may not be good for brain development. But you've already seen that acetaminophen damages brain development in laboratory rats and mice. In addition, acetaminophen is connected to a particular molecule called arachidonic acid,[259] and autism is connected with the same molecule.[260] But you've already seen many factors connected with acetaminophen and autism. Similarly, studies in laboratory animals indicated that acetaminophen is connected to damage to cells in a part of the brain called the cortex,[261] and autism is connected with damage to the same part of the brain.[262] Further, acetaminophen can adversely affect a cellular process called calcium signaling,[263] and that process is known to be important in autism.[264]

The most up-to-date list we have published in the peer-reviewed literature so far has only twenty-two lines of evidence,[265] with two additional lines of evidence published separately,[266] bringing the total to twenty-four lines of evidence. Our current list, shown in Appendix B, describes thirty lines of evidence that, taken all together, allow us to conclude that acetaminophen plays a critical role in the induction of autism. We or somebody else will eventually get the current and complete list of evidence published in the peer-reviewed literature, although it's already available for anyone to read.[267] It's important to note here that the evidence is *actually published*, but a list putting it all together is not yet published. The fact that a more current list of evidence is not yet in the peer-reviewed literature is not for a lack of trying.

We thought that a manuscript describing an updated list of evidence might fit into the journal *Paediatric and Neonatal Pain*, published by Wiley publishers, because they had previously invited me to serve as an anonymous reviewer for another manuscript dealing with the safety of acetaminophen for pediatric use. We submitted the manuscript to the journal on January 5, 2025. On January 20, 2025, Elaine Boyle, the editor-in-chief of the journal, informed me that the manuscript was not "a good fit with our journal's aims and content." The associate editor, Mats Eriksson, told us that the "format is not in line with the types of manuscripts we publish." We then submitted the manuscript to the *British Journal of Clinical Pharmacology*, another Wiley journal, on January 22, 2025. On January 28, the manuscript was rejected, with ten editors signing the rejection letter. The full extent of the rejection, with no peer reviews provided, was as follows:

> I am sorry but after assessment, I am unable to accept the paper for publication in the *British Journal of Clinical Pharmacology*. As you can appreciate, the Journal receives more manuscripts than can be published and so there has to be some degree of prioritization.

On January 29, 2025, we submitted the manuscript to *Advances in Pharmacological and Pharmaceutical Sciences*, yet another Wiley journal. Almost ten months later, on November 24, 2025, the manuscript was rejected by anonymous editors, again with no peer reviews. This time, the only comment was "After careful assessment, we have made the decision not to consider your manuscript for publication in *Advances in Pharmacological and Pharmaceutical Sciences*." The form letter did suggest we pick another journal from Wiley publishers, but we had already spent almost eleven months trying to get the article published in Wiley journals, and we had yet to receive an actual review. In addition, I think we have gone through every one of the journals that Wiley publishes that covers material related somehow to the connection between acetaminophen and autism. They do have one other journal that covers the topic of autism, but the editor of that journal is known to be dogmatically opposed to the idea that the environment contributes to the induction of autism. I expect we would be wasting even more time if we tried that journal. We have literally tried every journal from that publisher that we thought might be interested in the topic. None were.

A central point of our manuscript with the updated list of evidence was to question the wisdom of intentionally exposing babies to a particular series of vaccines that is so disturbing that the babies require three doses of acetaminophen in a single day. Use of the meningococcal group B (MenB) vaccine, which became available in 2014,[268] has spread rapidly with a protocol that includes three doses of acetaminophen that "should be given" each time the vaccine is administered.[269] Some countries, including Israel, the UK, Australia, New Zealand, and Canada[270] (but not the United States) are now recommending this protocol at two, four, and sometimes twelve months of life.[271] Based on the evidence you've seen in this book, it's very reasonable to question that practice. Is the treatment of this manuscript from Wiley's editors unethical? They sat on the paper for almost eleven months without getting it reviewed, as far as we know. That's certainly not normal. They did not obtain any reviews, as far as we know, even though they cover this topic in their journals. That's not normal either, at least for most of the journals we selected. The bottom line is that this irregular treatment might or might not reflect unethical behavior. We don't know what went on behind the scenes, and although their behavior was unusual, they didn't leave any firm evidence of unethical behavior.

One way for us to give this paper a better chance in the peer-review process is to tone down the title so that the take-home message is less clear. The current, very clear title is

"Three Mandatory Doses of Acetaminophen During the First Months of Life with the MenB Vaccine: A Protocol for the Induction of Autism Spectrum Disorder in Susceptible Individuals"

That title was selected when we already had a good idea that information about the connection between acetaminophen and autism would go public. We knew that, despite a vast amount of evidence in that paper, the title might be too blunt to be palatable for many editors. On the other hand, a title that would be less clear to the public but would have a better chance of getting published is

"Potential Neurodevelopmental Impact of Meningococcal Antigen Exposure with Standard Analgesic Administration"

We can probably do even better, making the title completely obscure and perhaps adding a bit of flair:

"Quintessential Balance: The Yin and Yang of Analgesics in Pediatric Practice"

Besides toning down our rhetoric, which would make the paper more palatable to editors but less useful to readers, we could break the paper up, separating our current list of evidence from our observations about the dangers of recommending procedures that require more acetaminophen exposure in the first months of life. We already have twenty-four lines of evidence published,[272] and that is quite sufficient to make the points that need to be made. However, it would be nice to have the list of all thirty published.

You may recall from chapter 1 that the first paper my group published describing the connection between acetaminophen and autism,[273] now more than a decade ago, was completely ignored. I suspect that the only good that paper ever did was to serve as a learning experience: if even *I* can't read the title and get the message, then the message will probably be lost.

I accept that getting something published in the scientific literature is better than getting nothing published there, but if the public doesn't understand the message, and nobody is willing to publicly broadcast the message, then the point of publishing in the scientific literature is questionable. Nevertheless, we will probably change the title and the abstract of our manuscript to soften the message, and may even cut sections out of the paper, focusing on the scientific evidence, especially the biochemical and genetic aspects of the evidence. Then we'll submit the manuscript to another journal and see what happens. We need to get the information into the peer-reviewed literature, with the goal of developing a trustworthy basis upon which public policy can be derived. At the same time, we will rely on other means, such as this book, to deliver the message to the public in a clear manner.

The next chapter will be the last chapter describing scientific evidence. In that chapter, I'll summarize scientific errors and address factors other than acetaminophen that may or may not cause autism spectrum disorders. In the course of examining that evidence, our final line of evidence will become apparent.

CHAPTER 8

LITTLE SCIENTIFIC BLUNDERS SUPPORT THE BIG SCIENTIFIC BLUNDERS

So far, we've covered the three major scientific blunders that led to the current and horrific state of widespread use of acetaminophen for babies and small children. We've also seen numerous other scientific errors. Here I'll summarize those errors, describe a few more, and then move on to discuss vaccinations and some other factors that have been implicated in the pandemic of autism.

The first error appeared half a century ago. In the 1970s, clinicians concluded that acetaminophen was safe for babies and children. The conclusion was repeated hundreds of times in the medical literature, becoming entrenched dogma. The underlying mistake was the incorrect assumption that babies and small children metabolize acetaminophen the same way adults do. This error was discussed in chapter 7.

In the mid 2010s, a scientific focus on autism induction during pregnancy began to coalesce. This focus persists to this day, despite the fact that the earliest data pointing at acetaminophen in the induction of autism involved early childhood. Current "rebuttals" to President Trump's warning about the dangers of acetaminophen[274] focus on pregnancy. The problem is that, based on available evidence, the main period of risk is at birth and early childhood. So, a focus on pregnancy misses most of the risk. This error was discussed in chapter 5.

Also in the mid 2010s, the error of treating interacting variables as confounding factors was introduced and has subsequently become entrenched in scientific practice. Current "rebuttals" to President Trump's warning about the dangers of acetaminophen[275] are completely derailed by this error. This is the most blinding error and was discussed in chapter 3.

Those are the *Big Three* errors, all involving what I believe to be honest scientific mistakes. These errors have persisted for over a decade. Next in my roundup of errors, I will lump a collection of missteps in scientific thinking into a category I call *Oops, That Was Wrong*. Most scientists will immediately recognize the errors in this category as wrong, but nevertheless they still do sneak into the discussion occasionally. I see much more of these errors on social media than anywhere else. Like the Big Three, the Oops, That Was Wrong errors are also technical mistakes. But these errors haven't affected scientific thinking as much as the Big Three. A few of these errors are listed here.

- **Acetaminophen has been used for so long and is used so much that if it really caused autism, we'd see autism universally by now.** This technical error fails to recognize that some individuals are more susceptible than others.

- **Correlation doesn't equal causation, so therefore you can't say acetaminophen causes autism.** This error fails to recognize that correlations (associations) may or may not be due to causation, and it fails to recognize the vast amount of data pointing toward a causal relationship between acetaminophen exposure and autism.

- **Blaming autism on one factor is oversimplifying a complex issue. A simple chemical alone can't explain something as multifactorial as autism.** This error fails to recognize that susceptibility to injury is a complex factor, affected by many things. Susceptibility makes autism complex, but it does not logically follow that a single trigger cannot be responsible. Further, this error fails to recognize the similarities between fetal alcohol spectrum disorder and autism spectrum disorder.

- **Studies in laboratory animals are not reliable and should not be used as a basis for any conclusions.** This error fails to recognize that numerous studies in laboratory animal models all point toward acetaminophen as a potent toxin for brain development, and that the results in animal models support clinical observations and multiple associations between acetaminophen and autism in humans.

- **Studies in siblings and twins prove that autism is genetic, so acetaminophen can't cause autism.** This error fails to recognize the scientific concept of "missing heritability." When something looks heritable but the genes responsible can't be found, then the best explanation involves complex environmental interactions with genetics.[276] The many associations between environmental factors and the prevalence of autism demonstrate that the principle of missing heritability applies to autism. The changing prevalence of autism and lack of autism in certain areas of the world also prove that autism is not genetic in the classical sense.

- **Fevers are associated with autism; therefore, fevers must be blocked.** This error fails to recognize that overwhelming evidence demonstrates that the treatment of fever with acetaminophen, not the fever itself, induces autism. Fevers are associated with treatment of fevers, and since treatment of fevers with acetaminophen induces autism in susceptible individuals, then fevers will be associated with autism. If A is associated with B, and B is associated with C, then A is likely to be associated with C.

- **My child was never exposed to acetaminophen, and he has autism; so acetaminophen doesn't cause autism.** This error is the classic anecdote fallacy. An exception does not make a rule. At the same time, many children are exposed to acetaminophen during labor and delivery or during a circumcision procedure, and their parents are unaware of it.

At this point in our roundup of errors, I'll shift my focus to errors more related to the scientific method than to specific technical errors. An error that I describe several times in the pages of this book can be called *Science Bashing*. Sometimes "science word salad" is used. The words *sound* like science, but they actually don't make any sense when you dissect out the text. Numerous examples include "isn't high-quality," "biased," "poorly designed," and "didn't address confounding factors." Any of these labels can be accurate if a study really *is* of poor quality, but the key in a scientific evaluation is to be *specific*. For example, what exactly was the confounding factor that supposedly undermines the study? As we now know, if that factor was an interacting variable, then the study is not actually undermined. If somebody says a study was of poor quality, or not high enough quality to pass peer review, can they say exactly *why* it was not of high quality? If they can't, they are science bashing.

Science advances when an observation doesn't add up within our current understanding, and we try to make sense of the observation. Science doesn't advance when something happens that doesn't make sense and we invent numerous reasons to ignore the observation. The bashing of studies involving connections between circumcision with social dysfunction has been particularly aggressive in the scientific literature, as discussed in chapter 7. And in the introduction, you can see that sometimes science bashing happens even before a paper is published (see also Appendix A). Calling a study "outrageous, overstated, illogical, and unnecessarily inflammatory" (Appendix A) without any specific evidence supporting those claims is science bashing at its worst. In our paper published by the Korean Pediatric Society,[277] we examined scientific literature bashing the landmark study by Schultz.[278] You may recall that the study was the first to direct attention to the connection between acetaminophen and autism. If the study had been taken seriously, evidence available at that time would have raised enough concern to drive changes in clinical practice and, we predict, alleviate the burden of autism.[279] However, the study was labeled as "biased" and largely ignored for that reason. A detailed analysis of the Schultz study reveals no bias that would affect the conclusions of that study in any way.[280]

Another distinct error I have encountered on more than one occasion is the *Scientific Bluff*. Similar to a bluff used when playing cards, the scientist claims to have some bit of evidence that he or she does not, in fact, have. We saw this type of error twice in chapter 7. One expert for the *European Journal of Pediatrics* claimed that a study had provided "reassuring" evidence that acetaminophen was safe for neurodevelopment. Fortunately, the journal's editor, Peter de Winter, recognized the bluff and called it. No such evidence exists. Another expert, this time working for the influential journal *Pediatrics*, claimed the existence of an imaginary phenomenon, "enhanced infantile sulfation," that protects babies from acetaminophen (chapter 7). Unfortunately, despite the fact that the scientific literature clearly indicates that no such phenomenon exists, the editor, Alex Kemper, allowed the paper to be rejected based at least in part on that bluff.

Another error, the *Error of Infinite Investigation*, is something that is strongly incentivized in the scientific community. We scientists need to keep our labs running, which means we need some reason to continue our research even if we really should sit back and call it good. This is not to say that we will ever know everything, but sometimes we know more than enough to make public health policy recommendations. If we don't need to know anything more to set public health policy, why should we get more funding from a federal agency focused on public health? The end result of this system-wide problem is that we tend to avoid drawing firm conclusions in favor of seeking more evidence, a never-ending endeavor. In our systematic review of the literature, 63 percent of all articles addressing the safety of acetaminophen for brain development called for more studies to address unresolved issues.[281]

In chapter 5, again an anonymous reviewer from the journal *Pediatrics* protested that nobody has observed increased levels of the toxic metabolite of acetaminophen under certain conditions (Appendix D)—despite the fact that the biochemistry *dictates* that the metabolite will be present in higher amounts under those conditions. Further, studies aimed at observing that metabolite under those conditions would be exorbitantly expensive. In our work on acetaminophen prior to 2020, a key administrator at the National Institutes of Health told us there was no point in applying for funding to examine the safety of pediatric acetaminophen use. This person, the scientific review officer (SRO) in charge of the study section on pediatric analgesic safety, had absolute control over funding priorities. There was no appeal process. (Funding priorities are not dictated by scientists; they are dictated by government administration.) The anonymous reviewer for *Pediatrics*, then, was demanding a study that could never be done unless I could find private donors to fund the work. And, as pointed out in chapter 5, if the study had been done, questions would still have remained.

In essence, the Error of Infinite Investigation is a type of *straw man fallacy*. It draws attention away from what we *do* know by highlighting what we *don't* know. We must examine the unknowns on the frontiers of science, or else science can't proceed. But we must also examine what we do know, especially when public health is at stake.

Another error can be called *Destroy the Weak Link*, and it abounds in the media. This error usually involves placing too much weight on a single bit of evidence and then focusing on the weaknesses of that particular bit of evidence. This is related to the Error of Infinite Investigation but is clearly distinct. Destroy the Weak Link focuses on the *weaknesses* of one study, whereas the error of infinite investigation focuses on *unknowns*. Destroy the Weak Link is quite different from Science Bashing, because destroying the weak link involves pointing out *legitimate* weaknesses in a study.

One of our main bits of evidence targeted thus far with this approach is the connection between circumcision and social dysfunction, discussed in chapter 7. (This evidence is also targeted by Science Bashing, as noted above.) The media has largely focused on the Danish study,[282] completely missing supporting work such as the US study on circumcision[283] and the study of acetaminophen exposure at the time of birth.[284] The media also doesn't realize that knowledge from the field of pharmacology has a bearing on the issue. Further, the general body of knowledge described in this book supports the connection between acetaminophen and autism. The Destroy the Weak Link strategy goes for what might be erroneously considered a "weak link" in a chain of logic. But it's rare for science to depend upon a single chain of logic. Rather, science is held together by multiple lines of evidence, some weaker and some stronger from a science perspective, but all pointing to the same conclusion.

The prevalence of autism in the Amish, discussed in chapter 4, has been another topic upon which the media has pounced. As discussed in that chapter, it's easy to find an expert who can look at one survey study from the Amish community and tell you it's not very reliable. It's not so easy to find an expert who can—based on a broad knowledge of the field—tell you that, yeah, those data from the Amish aren't very reliable—but data from several other places *are*. The list of errors we have described so far is as follows:

- The Big Three: initial approval of safety, focus on pregnancy, and confusing interacting variables with confounding factors
- Oops, That was Wrong: a sampling of less influential technical errors
- Science Bashing
- The Scientific Bluff
- The Error of Infinite Investigation
- Destroy the Weak Link

These errors, which do unfortunately creep into the scientific literature, can be easily identified. Other scientific errors can be more subtle.

As discussed briefly in chapter 5, it's not *absolutely proven* that the toxic intermediate of acetaminophen reacts directly with brain tissue. But it made sense based on everything we knew. That is, until a study from Mitchell McGill's laboratory at the University of Arkansas Children's Hospital in Little Rock found that the toxic metabolite of acetaminophen does NOT react with the adult mouse brain—even at doses that cause liver injury.[285] The study was published in 2025. McGill's laboratory concluded that the toxic metabolite of acetaminophen "is unlikely to contribute to the pathophysiology of neurodevelopmental disorders." This is a science-sounding way to say that, based on their data in mice, the authors conclude that it is unlikely that the toxic metabolite of acetaminophen is involved in the induction of autism, among other things, in humans. McGill uses the word "implausible" in the conclusions of the paper. He's suggesting that the whole idea of acetaminophen forming a toxic metabolite and then leading to autism is implausible. By definition, *implausible* means something unreasonable or improbable.

But all the evidence we have tells us that the toxic metabolite of acetaminophen forms *in the presence of oxidative stress*, and we know that oxidative stress is a factor that is associated with autism. So, what's going on? It turns out that the data published by McGill in 2025 is not the whole story.

In August 2022, I reached out to Laura James, who works with McGill at Arkansas Children's Hospital. The idea was to look for signs of acetaminophen toxicity in the *newborn rat* brain. We treated our newborn rats with acetaminophen, collected tissue, and sent the tissue to the University of Arkansas. Samples arrived in the McGill laboratory on November 15, 2022. On January 30, 2023, Larry Parker, the technician working on the project, provided me with a "rough guess" of the amount of toxic acetaminophen metabolite in the brains of our laboratory animals. The result was preliminary, which meant we needed to do more work.

Then our laboratory got busy working on the behavior of the rats when they grew up. It was complicated, extremely time-consuming, and literally took years. The thinking was that the data from Arkansas were great, but we would have to come back to it later. We lost touch with the McGill laboratory.

One of our brilliant undergraduate students saw the McGill paper published in 2025, showing no formation of toxic metabolite in the brain.[286] This was the first time we knew that McGill had continued working on the project without us. This is unusual in the field of science. Usually, if you start a collaboration with somebody, you don't continue the project without at least informing them. McGill's group had changed a number of things in their protocol compared to the early experiments we did with them. First, they changed from *rats* to *mice*. Then they switched from *newborns* to *adults*, and finally they *increased the dose* by twofold but *reduced the number of doses* of acetaminophen from four to only one.

We already knew from work in other laboratories that one dose of acetaminophen did not cause any long-term neurological damage in mice, whereas two doses of acetaminophen does cause damage in mice.[287] We also knew that *adult* mice are resistant to acetaminophen-mediated brain damage, but *newborns* are not.[288] We also knew that acetaminophen kills brain cells in adult rats.[289] That much was known from other labs. So, the experiment McGill's lab had done—administering a single dose of acetaminophen to adult mice rather than four doses to newborn rats—was designed quite well to *not* find damage from acetaminophen.

The options my laboratory had were limited. It seemed unlikely to me that additional work with McGill's laboratory would be possible—particularly since he had published a result that he and I both knew did not reflect the preliminary data we had collected together. McGill's group was the only group in the world I could find that ran this test. I had considerable experience with the necessary analytical instrumentation, but I knew it would take me the better part of a year to figure everything out. Since we already knew that acetaminophen kills rat brain cells, and we were finding similar results in our current studies, maybe this was just not important enough? Maybe we were going to have to abandon the project showing direct acetaminophen toxicity in the brain?

Fortunately, an organization called Children's Health Defense offered to help. They knew of an analytical laboratory that was willing to get the test up and running, and they were willing to fund the analytical laboratory. The analytical laboratory, Health Research Institute (HRI) in Fairfield, Iowa, agreed to collaborate with us. In August 2025, three years after we had begun the collaboration with McGill in Arkansas, we were up and running again. Again, we treated our newborn rats with four doses of acetaminophen, collected the tissues, and sent them off to the laboratory. The preliminary results came back on November 30, 2025. The results are shown in Figure 8.1. For comparison, the preliminary results from McGill's laboratory and the results that McGill published are also shown. You'll notice that the amount of acetaminophen toxicity measured in the rat brain is very close to the minimum amount associated with cases of acute liver damage in humans.[290] That's the dotted line in the figure. At least that's the preliminary result in the newborn rats with four doses of acetaminophen, but clearly not the results published using adult mice with a single dose of acetaminophen.

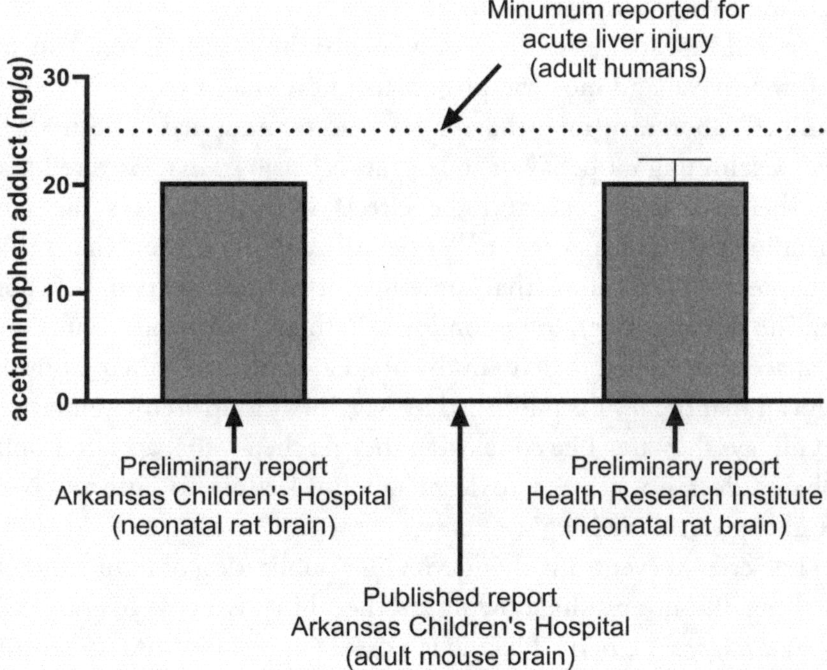

Figure 8.1: The results of acetaminophen binding directly to brain tissue in three different experiments. When acetaminophen binds to the brain, it's called an *adduct*. The presence of the adduct indicates that the toxic metabolite of acetaminophen was formed and reacted with a nearby protein, causing damage. McGill's published result found that the toxic metabolite was "absent in the mouse brain,"[291] indicated by zero on the graph. The dotted line shows the minimum amount of acetaminophen binding found in patients with acetaminophen-induced liver damage.[292] I know that this is not a science book, but I need to add this just in case a scientist wonders where I got the numbers from: I had to do some conversions between things called ng/g and nmol/ml, and to do that I assumed that the density of the brain tissue was 1.0 g/ml. My apologies for this brief digression into science jargon.

Keep in mind that the results in newborn rats shown in Figure 8.1 are "preliminary." That means we want to do the experiment again before publishing any results in the peer-reviewed literature. And repeat it again, and again, if that is what it takes to make sure everything is reproducible. Because they are preliminary, the results don't tell us exactly how much toxic metabolite is there—but it's definitely there. The exact amount is preliminary, but the fact that it's there is obvious.

Imagine counting a bunch of cows in a field. If you do it quickly to get a rough guess, you won't be sure exactly how many cows are in the field. But you can be sure there *are* cows. If you counted two hundred cows quickly, you can be sure there are more than zero cows, even if you don't know the exact number. Counting cows is sort of like analytical chemistry in that way. The toxic metabolite of acetaminophen was definitely in the newborn rats' brains.

We will repeat the results . . . as many times as it takes. You may recall that we abandoned the project for more than two years while we examined laboratory rat behavior after acetaminophen exposure. The reason examining rat behavior took years is because we discovered many never-before-seen effects, and the effects were not always the same, depending on the experiment.[293] I described our work on that project in chapter 5. The point is that sometimes it can take years to nail down scientific results enough to be confident in them. So far, our results looking at acetaminophen toxicity in the brain are only preliminary, but they do mean that the results published by McGill's group aren't as incisive as McGill says they are. The conclusion they reached—that acetaminophen probably does not form a toxic metabolite leading to autism—is not plausible given everything we know.

This series of events involving McGill's publication of results showing one thing, but not results showing another thing, is a type of error called *Cherry-Picking*.[294] Cherry-Picking is a classic sign of bias and undermines the progress of science. Years before doing any experiments looking at acetaminophen and the brain, McGill had concluded[295] that "In fact, there is strong evidence that (autism), in particular, is driven by genetics,[296] so exposure to (acetaminophen) or other xenobiotics may not be important." (A *xenobiotic* generally refers to toxins and drugs.) McGill has also worked as an "expert witness" defending Kenvue, the manufacturer of acetaminophen,[297] when they were sued by parents who believed their child's autism was due to acetaminophen exposure.

It is possible that McGill is biased.

Next, I'll consider other factors that people have proposed to be responsible for the induction of autism. I think that many people reading this book will be interested in vaccinations, and that is the most complex in terms of what we know. So, I'll do that one first.

The information about vaccines and autism falls into a spectrum between wrong, misleading, and absolutely correct.

I have an appreciation for how vaccines are intended to work. One of my very first science projects, conducted while I was an undergraduate in the 1980s, involved vaccinating laboratory mice to create different types of antibodies that we needed for research. In graduate school I vaccinated more mice, creating yet more antibodies that we needed for research. None of that was ever published, to my knowledge. As a faculty member at Duke University, I did finally manage to publish some of my work looking at the effect of vaccination on laboratory animals,[298] as well as some work on how the human immune system responds to vaccine-like exposures.[299] Because I worked at Duke in the Department of Surgery, which had one of the premier vaccine institutes in the country, I heard countless lectures on vaccine development. I gained a strong appreciation for how viruses avoid the immune responses to vaccines, making vaccination for some viruses, particularly HIV, extremely challenging.

All that being said, I have not spent 20,000 hours studying vaccines. I don't have detailed knowledge of the components in the vaccines, detailed knowledge regarding the reasons for giving the vaccinations, or detailed knowledge of the methods used to develop and manufacture commercial vaccines. I am not an expert in infectious diseases or how to prevent such diseases. Many scientists have incredible knowledge about one subject and then make sophomoric blunders when discussing another topic. I will avoid that.

I am very confident about some facts regarding vaccines as they relate to acetaminophen and autism. For example, if a seventeen-month-old toddler gets a vaccination and has a severe reaction to that vaccination, they are likely experiencing oxidative stress, which increases their sensitivity to acetaminophen-mediated injury. This reaction will then induce the parents to treat the toddler with acetaminophen, which may injure the brain, leading to autism.

Examples of parents who administered acetaminophen to their children because of adverse reactions to vaccines are presented in chapter 5. The children regressed into autism following that treatment for reasons that we understand. Stephen Schultz, the physician who performed the landmark study of 2008, provides yet another example. His son, Nathan, regressed into autism following administration of multiple doses of acetaminophen following an adverse reaction to a vaccination. That story is publicly available.[300] The bottom line is that we should not ignore the beliefs of a very large percentage of the population who have a child with autism. They have a rational basis for concluding that vaccination was involved in their child's autism.

The question often asked is, did vaccination "cause" the autism in these particular cases? If the answer must be either yes or no, then I think that the most correct answer would probably be yes—but that answer would be misleading. The reasonable answer is that the vaccine (or often multiple vaccines administered simultaneously) induced a reaction that (a) caused oxidative stress and increased susceptibility to acetaminophen-mediated injury, and (b) caused the physicians and parents to administer acetaminophen. The acetaminophen then triggered autism. In this case, asking a yes or no question misses not just the nuance of the scientific evidence, it misses the broad strokes of the scientific evidence. Here it's important to review Equation 4.1.

Equation 4.1: acetaminophen exposure + susceptibility = autism

Based on that equation, the role of vaccination in the induction of autism is not critical or absolutely required. If we avoid acetaminophen, we don't need to worry about autism—regardless of what reaction a baby or child has to vaccination. At the same time, extensive evidence points toward the induction of chronic immune dysregulation by repeated vaccinations in some individuals. This model of chronic immune dysregulation relates to Nobel Prize–winning research dealing with anaphylaxis that was conducted over a century ago by Charles Richet, professor of physiology in the Medical Faculty at Paris. More recent work by the renowned immunologist Yehuda Shoenfeld, professor of clinical immunology at Tel

Aviv University, also addresses this issue. A schematic diagram showing the multiple potential connections between vaccination and autism is shown in Figure 8.2.

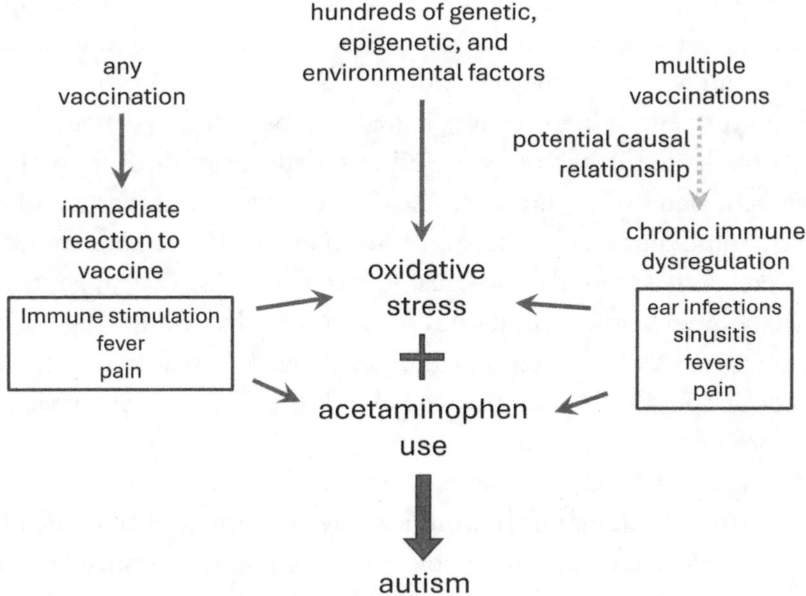

Figure 8.2: The potential roles of vaccination in the induction of autism.

I will emphasize again to the reader that I am aware that I'm not an expert on vaccines, except when it comes to their connection with the induction of autism. I am an expert in some aspects of chronic immune dysregulation,[301] but I am not qualified to write a book on adverse reactions to vaccines, or even on the potential induction of chronic immune dysfunction that might be related to repeated vaccination. I do, however, recognize signs of Science Bashing when I see it, and I know a Science Bluff when I see it. Evidence pointing toward chronic immune dysregulation after repeated vaccination exists, and it is a breach of scientific ethics to ignore that evidence. For that reason, I have included that component in the diagram. However, I am not sufficiently knowledgeable to address the topic in detail, and I will leave it to the reader to find the scientific literature on this topic if they are so inclined.

That being said, I can conclude that it would be an error to attribute the currently high prevalence of autism to vaccines alone, in the absence of acetaminophen. I included some discussion of this issue in my team's

2017 paper on the connection between acetaminophen and autism.[302] Studies showing associations between cord blood acetaminophen and autism, or between circumcision and autism, cannot be reconciled with the idea that vaccines alone induce most cases of autism. Acetaminophen given at the time of birth and circumcision are not associated with vaccinations, and without association, there can be no causation.

From a scientific perspective, it makes sense that a particular drug, related to alcohol in many ways, induces a spectrum disorder with distinctive characteristics. But vaccines are a very, very diverse group of substances. Some contain live viruses, while others contain specific molecules that were cloned from disease-causing microbes. Vaccines in general do not share any chemical properties in common that would suggest that they might all lead to a particular type of neurological injury or other toxicity. It could be argued that the aluminum found in many vaccines is the problem. Some of the first connections between vaccination and autism were reported by parents in the 1960s and involved the DPT shot,[303] which contains an aluminum adjuvant. As described by Rimland in 2000, "Soon after my textbook on autism was published in 1964, I began to hear from other parents. Many parents told me that their children were normal until getting a triple vaccine—the DPT shot."[304] But the MMR vaccine, which contains no aluminum, was also associated with autism. As we know, Schultz showed that acetaminophen was the common denominator when it came to the MMR shot and autism.[305] Others have suggested that ethylmercury, used as a preservative in some vaccines, was causing autism,[306] but dramatic reductions in the amount of mercury in vaccines has not resulted in a decrease in the prevalence of autism. The bottom line from a science perspective is that it makes more sense for a specific drug to cause a specific disorder than it does for a collection of vastly different drugs to cause that disorder.

Some incisive studies have been conducted addressing the possibility that vaccination induces most cases of autism. Ghassan Abu Kuwaik, a developmental pediatrician at the University of Toronto, led a team probing the prevalence of autism in unvaccinated children.[307] He evaluated the prevalence of autism in younger siblings of children who had autism. He found that, among twelve unvaccinated children who had older siblings with autism, four were diagnosed with autism. The study authors concluded there was no difference between vaccinated and unvaccinated children and decided they had too few children to draw any firm con-clusions. This assessment was not exactly an error, per se, but the authors did miss the most important part of their findings. The best question to ask would be, *Is it possible that the prevalence of autism in the unvaccinated children was the same as it was in 1925, when the prevalence was very low?* In other words, would it be possible to find a group of children with one third having autism if indeed the prevalence of autism was only 1 in 2,500? Someone could argue that these are younger siblings of children with autism, so the chances that they will have autism are higher. But we should keep in mind that the hypothesis here is that *modern* vaccines (or the modern vaccine schedule) cause the high prevalence of autism today. If that hypothesis is correct, then few if any of those older siblings would have had autism if they were born before 1925, because they would not have received modern vaccines. In addition, as was discussed in chapter 4, autism was much less common prior to 1970. It turns out that the odds are vanishingly small that we could find a group of twelve children with 33 percent autism if the prevalence was really 1 in 2,500. The odds are less than 1 in 10,000.

Anjali Jain, a physician scientist working with the Lewin Group, and Craig Newschaffer, a professor in the Department of Epidemiology and Biostatistics at the Drexel University School of Public Health, also evaluated the prevalence of autism in unvaccinated children, focusing again on younger siblings of children who had autism. The prevalence of autism was over 8 percent in 269 unvaccinated children with an older sibling having autism. Again, the odds that a group of children would have such a high prevalence in 1925 is remote, less than a 1 in 10,000 chance. Although Jain and Newschaffer only looked at avoidance of the MMR vaccine, it is likely that the children probably did not receive other vaccines as well, due to concerns that vaccinations may have induced autism in the older sibling.

I am aware of some emerging data involving unvaccinated children with low or very low rates of autism. It is important not to dismiss obser-vations, breaching the scientific process. It is important to understand how low vaccination rates could be associated with a low prevalence of autism in some studies, but not others. First, parents who decide not to vaccinate due to concerns about immunological problems or autism are increasingly aware of the connection between acetaminophen and autism. The result is that some individuals receive *neither* vaccines *nor* acetaminophen, and we would expect those individuals to have very low levels of autism. Second, some children are unvaccinated because they are neglected. Neglect can also lead to a lack of any medical treatment, including administration of acetaminophen, again leading to a low prevalence of autism.

The view that repeated vaccination causes chronic immune dysregulation is also potentially important. If repeated vaccination causes chronic immune dysregulation, that would lead to correlations between vaccination rates and autism. Immune dysregulation causes chronic immune problems, some of which are often treated with acetaminophen (Figure 8.1). In addition, chronic immune dysregulation would lead to oxidative stress, which would increase susceptibility to acetaminophen-induced injury.

To summarize, available evidence is consistent with the conclusion that vaccines, in general, do not cause most cases of autism. It could be argued that vaccines cause a few cases of autism. It may be exceedingly difficult to prove or disprove such an idea. However, again, autism spectrum disorders have specific characteristics that suggest a common, underlying cause regardless of whether onset is before (infantile autism) or after (regressive autism) developmental milestones have been met. The fact that vaccines in general do not share any chemical features in common suggests that vaccines in general do not cause autism. The idea that vaccines induce some autism in the absence of acetaminophen adds needless complexity to the scientific model. By needless, we mean that the explanation for the observations is not improved by the added complexity. Occam's razor is an excellent guide in this situation, encouraging us to avoid needless complexity.[308] Again, as shown in Figure 8.1, I am not suggesting that vaccines do not increase susceptibility to acetaminophen-mediated injury, or that vaccines are not a common reason for the use of acetaminophen. To suggest such things would be to ignore the observations of parents, which has proven catastrophic for medical science in the past.[309]

Numerous other environmental toxins have been proposed to contribute to the induction of autism spectrum disorder. I favor the views of Good[310] and Shaw,[311] who independently have named several environmental toxins that can interact with acetaminophen, leading to an increased propensity for autism. These toxicants include antibiotics and the pesticide glyphosate. My group's conclusion is that a wide range of factors, including numerous environmental toxins and an unhealthy Western lifestyle, result in an increase in oxidative stress. That increase in oxidative stress, in turn, increases the propensity for an adverse reaction to acetaminophen, leading to autism.[312]

Finally, I will point out that it is *not* an error to hypothesize that vaccines or any other factor trigger autism. It is, however, an error to favor the hypothesis when other views have much more supporting evidence.

The errors described in this chapter are, for the most part, legitimate errors in scientific thinking. They are, in essence, technical errors. These sorts of errors are not unethical but rather are a normal part of the process by which humans learn. While these errors can mislead us and temporarily blind us to reality, they should not halt scientific progress if the scientific process is allowed to work the way it is designed to work. Unimpeded science reveals and rapidly weeds out technical errors. On the other hand, many of the events described in this book involve something altogether different. The *breach of scientific ethics*, not the technical errors described above, is crippling the scientific process. While I have included Science Bashing and the Science Bluff in my list of technical errors, these border on breaches of the ethics of science.

In the next chapter, I will discuss subversion of the scientific process. Numerous examples of such subversion were evident in the events I described in this book. The question is, why does such subversion happen?

EMOTION-BASED OBJECTIONS TO THE SCIENTIFIC EVIDENCE

Your emotions are the slaves to your thoughts, and you are the slave to your emotions. **—Elizabeth Gilbert**

In looking at the expert reviews from the most influential pediatric journal in the world (Appendixes C and D), we found the following reasons for rejecting our evidence-based conclusions:

- Six verifiably false statements. Three obvious examples are
    (i) All the evidence is from our research team. (<u>Reality:</u> *Less than 10 percent of the evidence is from our research team.*)
    (ii) Our suggested course of action is nothing new and is already standard practice (<u>Reality:</u> *We suggest, for example, elimination of acetaminophen use during labor and delivery, which is still very common.*)
    (iii) Our suggested course of action is impossible. (<u>Reality:</u> *It is difficult to imagine how the objection that our recommendations are already standard practice is reconcilable with the idea that our recommendations are impossible. In fact, many changes to current standards of practice can be made with no risk. (See chapter 11))*
- Three irrational arguments. An example is that the Danish circumcision study can't be considered as evidence because acetaminophen use was not measured in that study. (<u>Reality:</u> *This objection was raised despite the fact that acetaminophen is widely used for pain relief during circumcision,*

and the reviewer had no explanation for the association other than the involvement of acetaminophen in the induction of autism. Even if the study team had access to records of acetaminophen use in the hospital, it would not have been overly helpful. Parents are often told to use the drug when they get home, because the babies tend to be fussy for twelve to twenty-four hours after the procedure.)

- Two true and irrelevant assertions. For example, the conclusions are not widely accepted. (<u>Reality:</u> *This is a formal statement of a consensus bias.*)

- One baseless conjecture: maybe lots of other studies showing that acetaminophen is safe have been conducted but not published. (<u>Reality:</u> *Our team evaluated every study ever conducted that supposedly showed that acetaminophen is safe,*[313] *and none of those were valid. Further, studies evaluating the safety of acetaminophen for brain development have consistently shown the drug to be unsafe, as described in detail in this book.*)

- Insufficient references and self citations were frequently used, indicating poor quality work. (<u>Reality:</u> *The chief editor limited us to seven references after we had initially submitted a fully referenced manuscript.*)

- Disagreement with no valid justification. (<u>Reality:</u> *This is an example of the invincible ignorance fallacy.*)

I can see how, after reading this book and that litany of errors from the gatekeepers of the world's most influential pediatric journal above, one might wonder what in the world is going on. Where has our common sense gone? Has our cheese completely slipped off our cracker?

I believe I can answer that question. The problem is not a lack of intelligence. Scientists do not lack intelligence. In fact, their intelligence often works against them.[314]

Imagine a luxury cruise ship sailing for whatever part of the world we find most interesting. The ship will take us to places never before seen by any human, for us to explore. Getting a place on this cruise ship is difficult. Space is very limited. We must work hard to get there. It will take years of work, and most people will fail. If we are accepted, the intellectual rewards and the prestige are second to none. We'll be seen as an intellectual giant at parties and family gatherings, and we'll have the respect of our community and colleagues. And, if we do really well, we'll gain international recognition. If we are truly exceptional, we may even win a Nobel Prize.

We've spent the first half of our life working toward this goal, and we finally made it. Just imagine. We are a passenger on the ship, achieving something that 99 out of 100, or maybe 9,999 out of 10,000, never could. We are in a top-tier academic institution, conducting research that goes where no human has gone before.

But in the process of getting there, we have become aware that the ship only carries enough supplies for about 20 percent of the passengers—and the ship's administration expect the people on the ship to figure out how to sort that out. On top of that, the ship's administration is trying to recruit new people to join the ship, despite not having enough resources to support the people they already have. We've worked so hard to get our place on the ship that we ignore the inhumane, dog-eat-dog system. We fall victim to a sunk-cost fallacy, with our subconscious driving us to believe that our efforts are worthwhile. We turn a blind eye when a senior colleague is left at port after their supplies run out, with a big farewell party but then no thought afterward. They are left to regret that they spent so much time investing in a system that seems to no longer care that they ever existed.

The ship administrators have rules in place to govern how the resources should be sorted out—but nobody carefully monitors who plays by all the rules and who doesn't. Survival becomes paramount, and the system evolves. The best at survival learn just how much to bend or even break the rules to achieve the best outcome. It is survival at its finest. If we break the rules and thrive, we are smart. If we break the rules and the tribe catches us, we are humiliated and ostracized.

Human nature becomes a significant factor as our subconscious grapples with the prospect of losing our place on the ship. If we do lose our place, we lose our status in our community and possibly even the ability to support our family. A lifetime of work is lost, and we must start over with something unknown and uncertain. A survival state dominates our mind, as our subconscious drives emotionally violent reactions to threats, and we are unable to accept ideas that our subconscious knows are likely to be ridiculed by our peers.

Our emotions and our subconscious begin to dominate our thinking. When somebody suggests, for example, that the field of immunology is failing because it has stood by helplessly as a plague of chronic immune disease engulfs the country, we recoil in anger: We are a renowned immunologist. We are NOT a failure. We have discovered many wonderful things. Our students and our staff adore us. We have prestigious titles in a prestigious place. If the people around us worship us, we cannot be a failure. That's impossible!

Emotions have us under their control, in a state of "threat rigidity."[315] In that state, our ability to think creatively and objectively is limited. Our brain wants to defend our status in our tribe in order to survive. Our brain is still in control, but it's not the rational part of our brain.

Can you imagine yourself in that position?

This is the practice of modern science. Intense competition combined with peer review is a recipe for catastrophic paralysis. Separately, these two ideas are wonderful. Competition breeds innovation and excellence. Peer review and collaboration work wonders, since the whole is much greater than the sum of the parts. But the combination is toxic. Although government administration determines what topics are funded, our peers are the ones that determine whether we get funding. Protecting our place on the precious ship becomes more important than the scientific process. We must stick with the dominant group and try to connect with that group in a way that secures our place on the ship. The incredible power of consensus bias, demonstrated in numerous scientific studies,[316] is amplified as our survival depends on it.

This emotional takeover is not a conscious decision. The drive to survive is part of our human nature. If we did not have that nature, we would not have survived as a species. The existential threat to our position in our tribe creates an emotional compromise, which drives cognitive dissonance. We are now capable of believing that the field of immunology has made wonderful progress, despite the fact that our country has become overrun by chronic immune disease. We can believe that our research in the field of autism is laudable, despite the fact that 3 percent of the population are now affected by the disorder, and we have no idea why.

Can you imagine being in that position? I don't have to imagine. I've known many people in that position who are wonderful, kind people. It's real. In a nutshell, the existential crisis created by peer review and intense competition has biological consequences that are unavoidable.[317] Being smart makes it worse.[318]

Misinformation, miscalculation, and misinterpretation plague the science connecting acetaminophen and autism, as you have seen. However, these scientific problems are likely only a symptom of underlying causes. Given the long history of acetaminophen use in the pediatric population and the extensive damage already incurred as a result, most if not all stakeholders are potentially faced with a variety of hurdles (Figure 9.1), many of which reinforce one another. For example, conflicts of interest and emotional compromises face scientists and clinicians whose careers and reputations may be damaged by acknowledging the role of acetaminophen as a critical trigger of autism. How would it look to our peers and to the world if we worked on something for decades and completely missed the answer? The pressure from these factors could lead to denial. We simply can't be wrong. It cannot be this simple. This concept will be explored further in the next chapter.

The result is continued support of acetaminophen use, with pressure to avoid publishing work clearly describing the connection between acetaminophen and autism. These results could, in turn, make it more difficult for parents and parents-to-be to receive and act upon information regarding the role of acetaminophen in the induction of autism (Figure 9.1).

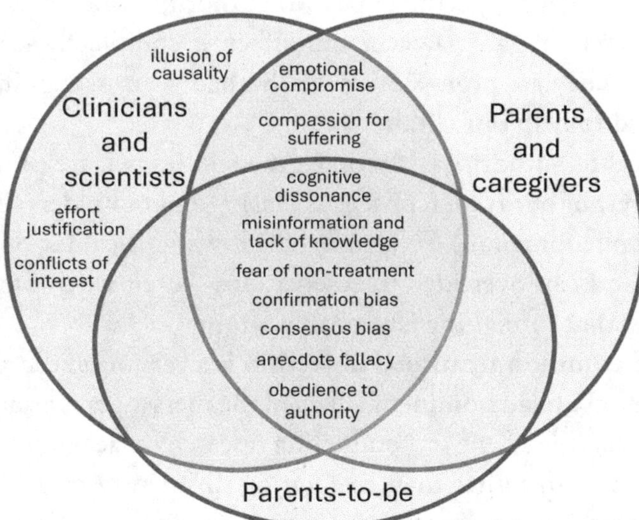

Figure 9.1: Potential nonscience factors affecting continued use of acetaminophen before the age of 6 years despite overwhelming evidence that the drug is a developmental neurotoxin involved in the induction of both infantile and regressive ASD. The list is not intended to be exhaustive, and some or even all factors may not necessarily affect all individuals. None of the factors affecting continued use of acetaminophen are evidence-based, but rather are aspects of human nature or results of aspects of human nature manifested in response to the current environment. This figure was published as part of a preprint that is available online.[319]

Studies of human cognition demonstrate that greater analytical ability *increases* rather than *decreases* susceptibility to confirmation biases.[320] What this means is that any improvement of analytical skills as a result of professional training in the fields of science and medicine could, inadvertently, amplify rather than alleviate difficulties overcoming the biases incurred during that training (Figure 9.1). Regardless of the underlying reasons, difficulty in overcoming long-held beliefs, despite abundant evidence undermining those beliefs, appears to be a significant factor promoting continued use of acetaminophen during early brain development. And the same difficulty in questioning long-held beliefs appears to drive breaches in scientific ethics. We have seen several examples of that in this book.

The conclusion that acetaminophen is a critical component in the induction of autism has elicited a number of emotional, irrational objections. I am grateful to colleagues who scour the web, finding these objections and summarizing them for me. I will provide a list here, along with the response to each objection.

The first and probably the most common irrational objection I encounter is the fallacy of the appeal to authority. Since medical organizations or pediatricians say acetaminophen is completely safe, then it must be safe. They are professionals who've had years of training specifically designed to help our children.

The appeal to authority is alluring. As a species, we are programmed to trust the rest of our tribe (consensus bias)[321] and our elders (obedience to authority phenomenon).[322] That behavior was critical for our survival as a species and can override our reservations—even when our subconscious knows that something is probably wrong.

A second common irrational objection is a version of the abusive ad hominem fallacy. In ad hominem fallacies, the messenger is attacked. The argument is that the scientists conducting research on acetaminophen are stigmatizing autism rather than celebrating diversity. According to this argument, scientists are literally producing hate speech against autistic children.

From a scientific perspective, this notion is preposterous. The identification of a chemical that induces autism during neurodevelopment means that autism is a chemically induced neurodevelopmental injury. Based on the evidence, autism is not, by definition, a part of a normal spectrum of neurodiversity any more than people with broken arms are a normal part of arm-movement diversity. If we want to be helpful as scientists, and we uncover the reason for an injury, we cannot label that injury as "normal," even in cases where injury is not severe.

A third irrational argument is another version of the abusive ad hominem fallacy. The argument is that the scientists conducting research on acetaminophen are blaming doctors and parents for autism. As you have seen in this book, I attribute the problem to several scientific errors coupled with emotion-driven breaches in the scientific process. I appreciate the fact that future work may uncover malicious behavior on the part of individuals seeking personal gain. Although such activity is not evident to me at the present time, I cannot rule it out. Regardless of the problems that have led to the current situation, it seems absurd to blame *caregivers* who are doing what they believe is right for their child. We have never assigned blame to those caregivers.

A fourth irrational objection involves the fallacy of guilt by association. Examples include "This sounds alarmist, so typical of today's times. These sorts of sensational news items should be ignored." Another example, this one related to the abusive ad hominem fallacy, is "That politician made the announcement, and nothing he says can be trusted. Everything he says should be ignored."

As a species, our brains are constantly trying to make sense of the world around us. Grouping things together and labeling them as all good or all bad is a powerful tool that our brain uses to sort out a complex world. The resulting guilt by association and similar biases creates tremendous barriers to rational thought. Our use of biases as a cognitive tool has helped us survive as a species. Unfortunately, as in this case, those same biases can cause us to miss vital information.

A fifth irrational argument I encounter is the invincible ignorance fallacy. In a nutshell, the entire idea that acetaminophen causes autism is ludicrous, regardless of the weight of evidence. A very animated example of this fallacy was produced by an expert reviewer for a scientific journal, shown in Appendix A.

A sixth irrational argument I frequently encounter is a specific non sequitur fallacy. A non sequitur fallacy involves a conclusion that does not follow from its given support. In this case, the fallacy is that, if it is this simple, it can't be correct. In fact, Occam's razor tells us that, as long as it explains the observations, simple is better.[323]

Another version of this irrational argument is that, if it *were* this simple, somebody would have figured it out by now. But no law or principle of science tells us that simpler things get figured out quicker. According to Max Planck, we may need to wait for senior scientists to grow old and die before new ideas are accepted. I agree with the legendary quantum physicist that the time it takes to figure things out does not depend on its complexity, but rather it depends on how much it offends individuals in power. With that in mind, it is not so surprising that the connection between acetaminophen and autism is not yet widely accepted. This will be explored further in the next chapter.

The bottom line is that the competitive systems we have created to manage and conduct science encourage our tribal instincts geared for survival, frequently leading to emotional and irrational behavior. Given human nature, it is only logical that such a competitive system, with survival always in question, would not bring out the best in our species. The challenge at the present time is to move science forward quickly despite the limitations of our system.

My favorite part about scientific research is no longer discovering something new that nobody has ever seen before. In my experience, the more exciting the discovery, the more slogging it will take to get it accepted by the scientific community, regardless of how convincing the evidence may be. Nicolaus Copernicus and Galileo Galilei, with their sun-centered model of our solar system, probably possess the grand prize for the greatest discovery with the most slogging involved. The redeeming part of science is that, sooner or later, reality will become clear. Hopefully, it won't take the scientific community centuries to figure out the connection between acetaminophen and autism. I pray that Copernicus and Galileo get to keep their prize.

CONFLICTS OF INTEREST IN THE MEDICAL AND SCIENTIFIC COMMUNITIES: TOO MUCH WAS BET TOO SOON

I am not a gambler. My ever-patient wife and I love to play poker, a card game that is frequently used for gambling. We play for tokens (called chips) that we inherited from her father. The idea is that you don't know what the other person's cards are, and you bet your chips on whether you think your cards, called your *hand*, are better than your neighbors'. One of my colleagues is quite good at the real game, which is played for money, not for chips that go back into a box at the end of each game.

A good way to look at the science surrounding acetaminophen and autism is through the lens of a poker game.

Prior to 2008, the scientific landscape related to autism was very, very different than it is now. The idea that acetaminophen triggers autism had not yet been proposed, but modern biomedical researchers had already placed their bets. The scientific consensus was that autism was innate, a complex mixture of environment and genetics, and not associated with any particular cause. The hypothesis that vaccines caused autism became popular, and the scientific community pushed back hard against that hypothesis, becoming further vested in the view that nothing in particular causes autism. The neurodiversity movement emerged and grew, further supporting the idea that autism was innate, without a cause. The neurodiversity view was even published in the scientific literature as "truth" in

January 2009.[324] Figure 10.1 shows how the bets were laid. Much was at stake, most importantly the well-being of so many children. However, the reputations of scientists studying autism were also at stake. If indeed there was some specific factor that was causing autism, then they had all missed it. *The problem here is that, before anybody had ever even thought that acetaminophen might trigger autism, so much was already bet on the idea that nothing in particular causes autism that losing was unthinkable.*

What causes autism?
Before 2008

Something

Nothing in particular

The stakes
- Autism in 1% of children
- Scientific reputations
- Financial liability

Unknown cause
- ???
- ???
- ???
- ???
- ???

Misunderstanding
- Very complex
- Prevalence not increasing
- Genetic
- Naturally occurring neurodiversity
- Innate

Figure 10.1: A poker game analogy of the science of autism prior to 2008. Before anybody even thought of the idea that acetaminophen might trigger autism, so much was already bet on the idea that nothing in particular causes autism that losing was unthinkable.

Everything changed with the publication of the Schultz paper[325] that first proposed a connection between acetaminophen and autism in mid-2008. Hindsight revealed abundant evidence pointing toward acetaminophen as a trigger for autism,[326] and dramatic effects of acetaminophen on brain development were observed in laboratory animals by 2014.[327] Figure 10.2 shows how the bets were laid out by 2014. The stakes had gone up, with the prevalence of autism being higher than it was prior to 2008. The reputations of scientists studying autism were still at stake, and since these individuals serve as gatekeepers for the peer-reviewed scientific literature, shifting the thinking would be difficult. As Max Planck knew, problems with entrenched scientific thinking are to be expected. But the newly revealed opponent, acetaminophen, added an extra dimension to the stakes. As you know, based on the information I presented in this book, I am very strongly opposed to blaming physicians for the use of acetaminophen. It is not wrong to do what you are taught is best for the baby under your care. But physicians nonetheless tend to take it personally when someone suggests that the drug they administered has caused autism. Therefore, in the minds of many physicians, *their* reputations were also at stake.

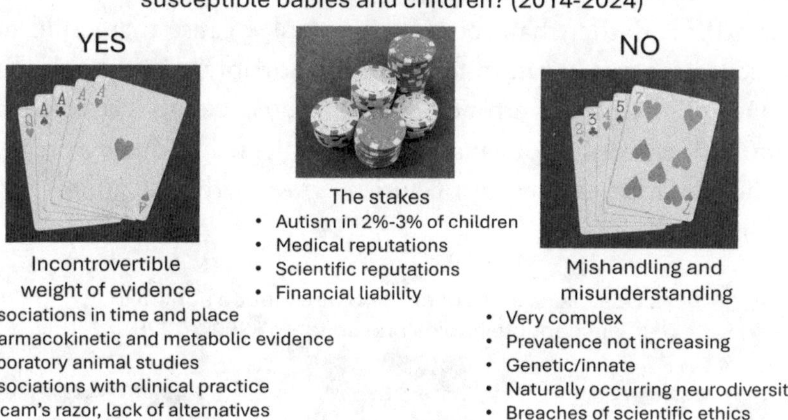

Figure 10.2: A poker game analogy of the science of acetaminophen and autism between 2014 and 2024.

Unlike an actual game of poker, everyone can see the cards—the evidence. To me as a scientist, the winner is evident. So much evidence supports the causal connection between acetaminophen and autism that no reasonable questions remain. The cards on the other side cannot win. The observation that autism is complex was incorrectly taken to mean that no single factor could cause autism. This seems logical at first glance, but the logic is not valid. Our first major paper on this topic explained clearly how autism could be complex AND still triggered by a single drug.[328] From a biological perspective, susceptibility to toxins is often a very complex issue. So, complexity in the susceptibility to acetaminophen-mediated toxicity is truly nothing particularly unusual or special. Unfortunately, the acceptance of complexity in the *cause* of autism led to incredible reductionism in scientific investigation, with different laboratories studying different aspects of the disorder. Uncovering greater and greater detail led to increased funding, which in turn promoted more reductionistic scientific work. Analysis of siblings and twins led to the erroneous conclusion, still believed by some, that autism is genetic. But as we saw in chapters 2 and 3, that is not a reasonable conclusion for many reasons.

Things changed dramatically in 2024 with the publication of the Drexel paper examining Swedish children and acetaminophen exposure during pregnancy.[329] Previous sibling control studies had been conducted, but none were considered so incisive. The arguments against acetaminophen as a cause of autism changed but were no more valid than before (Figure 10.3). Rather than focus on a complex cause that could not be identified, scientists began to focus on the horribly misinterpreted study from Drexel.[330] Given its erroneous conclusions, scientists concluded that acetaminophen could not cause autism—at least not during pregnancy. And the potential induction of autism *after* birth was almost entirely ignored.

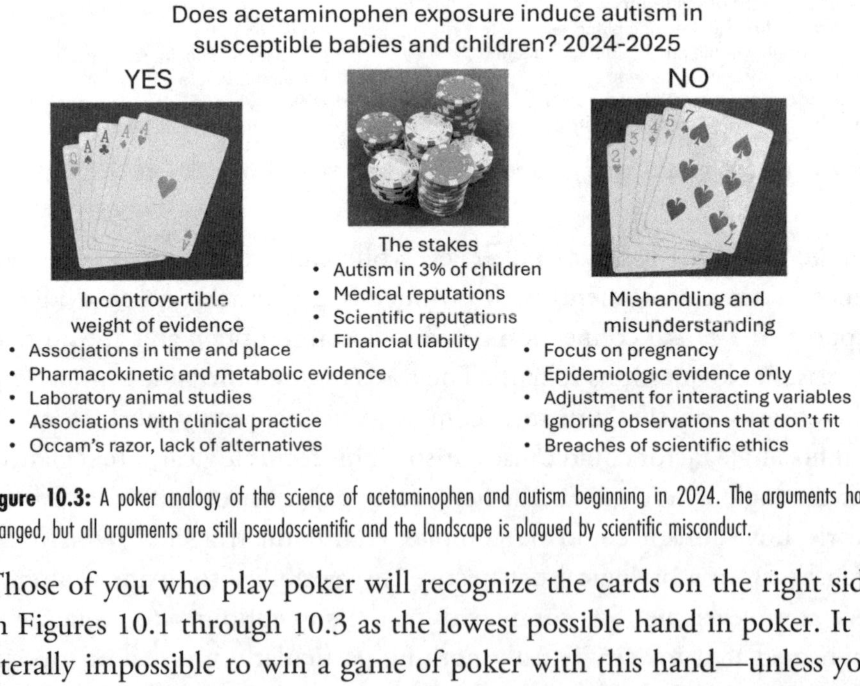

Figure 10.3: A poker analogy of the science of acetaminophen and autism beginning in 2024. The arguments have changed, but all arguments are still pseudoscientific and the landscape is plagued by scientific misconduct.

Those of you who play poker will recognize the cards on the right side in Figures 10.1 through 10.3 as the lowest possible hand in poker. It is literally impossible to win a game of poker with this hand—unless you bluff and the other players back down. You may also notice that I've picked a very good hand on the side of our conclusions, the left side of the figures. It's not the absolute best hand in the world—we don't have absolute proof—but that hand will win almost every time. The important thing to keep in mind here is that we can see the other player's cards.

They don't have anything. They can attempt a bluff, but it can't work because the cards are already visible. The only possible way they can win is to prevent us from showing *our* cards in the peer-reviewed literature. That does seem to be an effective maneuver, but it constitutes scientific misconduct and will eventually fail.

The problem for medical science is that it's hard to give up on that losing hand. The stakes are too high, causing extensive conflicts of interest and subconscious emotional compromise.

In late 2025, one of our published papers[331] was being "investigated" because a "concerned reader" said that it is flawed and will cause public harm. An editor from the publisher MDPI handled everything. Here I will provide the entirety of the complaint as transmitted to us by the investigating editor, along with my responses to the editor. I think many readers of this book will be surprised at the simplicity of the discussion between me and the editor. The editor's unedited description of the reader's complaint is listed in four points and is shown in bold.

Concerned Reader complaint 1. The correspondent notes that the article makes strong causal assertions (e.g., that acetaminophen "causes" or is "likely responsible for" neurodevelopmental disorders) but suggests the evidence presented relies heavily on animal studies, ecological correlations, and observational data susceptible to confounding. It is noted that some large human studies (including sibling-comparison analyses) that found no causal relationship after controlling for confounders are not addressed. The concern is that the level of certainty expressed may not align with the existing evidence base.

Response to Concerned Reader complaint 1: The reader has noted "large human studies" (including sibling-comparison analyses) that found no causal relationship. One particular study in question, and an excellent representation of other studies, was published in *JAMA* in 2024.[332] However, that study published in *JAMA* made rudimentary errors in their statistical calculations. In fact, the "sibling control" was not a sibling control per se, as frequently reported. Rather, the investigators corrected for over 30 different factors that are known to cause oxidative stress and interact with acetaminophen before adjusting for siblings.

(Oxidative stress interacts with acetaminophen to induce toxicity).[333] It is a known principle in statistics that interacting factors cannot be treated as confounding factors. Most importantly, we were able to fully recapitulate the results of the *JAMA* study, showing that their "correction" for confounding factors eliminated completely the real association that actually does exist.[334] Essentially, the *JAMA* work proved that acetaminophen is safe during pregnancy, but only if no risk factors that predispose children to adverse reactions to acetaminophen are present. This is, of course, not how they interpreted their results. Our proof of the error was published in 2024 after the *JAMA* study, although we had addressed the issue previously to some extent in an earlier article.[335]

Concerned Reader complaint 2. Representation of the Literature
A concern is raised that the review may treat citations that report only associations or weak correlations as robust causal evidence, while omitting or downplaying prominent methodologically rigorous studies showing null associations when confounding is addressed. The correspondent suggests this presents a selective view of the overall scientific evidence.

Response to Concerned Reader complaint 2: We are extremely clear that it is the weight of total evidence, not any one line of evidence in particular, that allows us to draw conclusions. Evidence includes numerous associations in space and time, more than a dozen studies in a variety of laboratory animal models, numerous lines of evidence from pharmacokinetic considerations, and independent associations during pregnancy, especially the peripartum period, and during early childhood. Again, we have published the mathematical proof that the "rigorous studies showing no associations" are fundamentally flawed. We have been able to recapitulate those results, and demonstrated the error.[336] Please do let us know if you need more information in this regard.

We do understand that many readers are confused about this issue, and we are working to correct that problem. It is a public health science communication problem.

Concerned Reader complaint 3. Adherence to Evidence Standards
The discussion is described as blending evidence from different

levels (e.g., rodent studies, human correlations, ecological trends, mechanistic speculation) without adequately differentiating their weight or validity in establishing causality, which may not align with established standards for evaluating epidemiological evidence.

Response to Concerned Reader complaint 3: Science is most clear when all evidence points toward the same conclusion. It would prove difficult and would indeed be almost arbitrary to weigh, for example, evidence showing that children with autism cannot safely metabolize acetaminophen with evidence showing that male laboratory animals are very sensitive to the drug during neurodevelopment or the evidence showing almost fourfold more autism with high levels of acetaminophen metabolites at the time of birth. The point is that all of the evidence points to the same conclusion.

Most scientists who do not understand the connection between autism and acetaminophen are indeed focused on epidemiology, and as mentioned earlier, those studies are fundamentally flawed. Again, we have proven this in statistical/math terms. If the interpretation of the epidemiological evidence does not match numerous other lines of evidence from other fields of science, then the interpretation of that epidemiological evidence should be reconsidered. In reconsidering that evidence, we have established the underlying problem with the interpretation of that evidence. The proof is published in an MDPI journal.[337]

Concerned Reader complaint 4. Public Health Communication Given the public health implications of the topic and the use of acetaminophen to treat fever during pregnancy, the correspondent expresses concern that the paper's definitive conclusions, if not fully supported by high-quality evidence, could influence patient behavior in potentially harmful ways.

Response to Concerned Reader complaint 4: A careful reading of the article in *Children*[338] will show that pregnancy is probably the time period of least concern. The highest risk for acetaminophen exposure likely occurs at the time of birth, in the immediate post-peripartum period. We agree that this topic is of immense public health importance, with 3 percent of all individuals having autism at the present time. The data from Johns

Hopkins regarding the association between acetaminophen and autism at the time of birth is especially concerning.[339]

The risk during pregnancy is probably only significant with high doses of acetaminophen used for chronic pain management. Again, we cannot be certain about risks during pregnancy, and the article in *Children* is quite clear about that. We also have an article published in the flagship journal for the Korean Pediatric Society that discusses the same issue.[340]

However, I should emphasize that the overriding public health concern here is the high prevalence of autism, and the clear link of that prevalence to acetaminophen exposure. Again, that link is based on more than two dozen lines of evidence. Importantly, contrary to the opinion of your concerned reader, there is no evidence to the contrary.

We do understand the public health problem caused by the fact that some supposedly rigorous epidemiologic studies are completely in error. It is not uncommon in science that it may take considerable amounts of time to correct an error once it has been propagated. This is indeed tragic.

Summary from editor: We would be grateful for any explanation or comments you can provide and look forward to your response by 10 December 2025. Thank you for your cooperation in this matter.

Response to the editor's summary: I am not opposed to addressing questions from misguided readers. Thank you for bringing this to our attention. If you have further questions, please let us know.

As you can see from the above correspondence, the "concerned reader" is dismissing the entirety of all available evidence because one study, not even focused on the period of highest risk, says no association exists between acetaminophen and autism. And, as I described in some detail (chapter 3), the conclusions of that study are entirely wrong because they adjusted for factors that interact with acetaminophen.

We speculate that the concerned reader was probably not an academic scientist. Even a relatively incompetent academic scientist would have found our more recent papers[341] that answered all his or her questions. We suspect also that this effort does not involve industry interference, since (a) such interference would leave a trail that would eventually be uncovered, (b) any successful attempt to have one of our older papers retracted would serve no purpose from a legal standpoint, and (c) such efforts are only likely to bring more attention to our work.

The language of the complaint sounds much more akin to those we hear from "social media experts," many with some degree of medical training. This may be a classic case of astronauts telling engineers how to build the spacecraft. Fortunately, after a few weeks of assessment, the editors dismissed the complaint and closed the investigation. Here again we see a wonderful example of people making a critical difference.

The complaint was rapidly dismissed by the editor. As I understand the situation, the editor offered the complaining party an opportunity to submit a commentary, and my research group would in turn be allowed to respond to that commentary. This is a wonderful way to bring important discussions into the public domain using the peer reviewed literature. Apparently, the complaining party declined to go on record with their complaint, so the idea was abandoned.

We are exceedingly grateful to the MDPI editors for their objectivity in this case. They went above and beyond their obligation, selecting our article[342] demonstrating the error of treating interacting variables as confounding factors (the primary error that misled the "concerned reader") as an *Editor's Choice* article. According to the publisher's website, "Editor's Choice articles are based on recommendations by the scientific editors of MDPI journals from around the world. Editors select a small number of articles recently published in the journal that they believe will be particularly interesting to readers, or important in the respective research area. The aim is to provide a snapshot of some of the most exciting work published in the various research areas of the journal."

Although the complaint against our article was dismissed, that complaint highlighted the fact that we failed to effectively communicate our work published in 2024 disproving the conclusions of the Drexel study. The failure is so complete that the Drexel study is carrying the day in major medical journals as of late 2025.[343] Here's the story of how we failed, at least up to this point.

After the Drexel study came out, we rapidly set the record straight. We published our paper[344] with a clean and clear mathematical proof that the conclusions of the Drexel study were entirely wrong. This is the title:

"Evaluating the Role of Susceptibility Inducing Cofactors and of Acetaminophen in the Etiology of Autism Spectrum Disorder"

This is an appropriate title for a science journal, but it will not communicate anything to the public. The abstract is clear enough, but most people don't make it that far. At this point, everything was going according to plan.

As soon as the paper was published, we sent out a press release that said something quite clear to the public. The supposedly ironclad work from Drexel "proving" that acetaminophen does not cause autism had been overturned. The press release started getting considerable attention, then, within twenty-four hours, the release *disappeared*, apparently after automated search engine bots detected something offensive.[345] We could not even find our own press release using Google. This was *not* the plan.

Then, as I have described (chapter 3), we failed in an attempt to bring attention to our work in 2025 due to unusual editorial activity at the journal *Environmental Health*. We also sent in a response to the horribly misguided article that appeared in *Nature* addressing this topic,[346] but did not hear back from the journal. That is typical for *Nature* and several of the most prominent medical journals. Publishing editorials is usually reserved for an inner circle of elites.

During our training as scientists in the field of biophysics and biochemistry, my colleagues and I were not taught much if anything about communicating science to the public. Maybe scientists working in more public-facing fields learn something about communication with the public, but we never did. I don't think such training would have helped, however. The public health science communication experts I worked with on this project have never seen anything like the situation we currently face. Two different professional marketing teams (Research Outreach and HUT 3, both based out of the UK) declined to help us, each after taking months to analyze the public relations landscape concerning our work. It was a learning experience for us all.

The scientific conclusions are still the same. And the conclusions are still obvious based on all available evidence. We are still certain without reasonable doubt about the following:

1. Exposure of susceptible babies and children to acetaminophen causes many if not most cases of autism spectrum disorder.
2. The best explanation for all available evidence is that the vast majority of all cases of autism, perhaps more than 90 percent, is triggered by exposure of susceptible babies and children to acetaminophen.
3. Susceptibility is caused by a complex array of genetic and environmental factors that lead to *oxidative stress*, a condition in which the body cannot keep up with the chemical burden it faces.
4. The time of greatest risk for brain development is during and immediately after birth. Risk diminishes within weeks and dissipates almost completely by the time the child reaches six years of age. Some risk during pregnancy seems likely, but with less risk and with less certainty of risk than at birth.
5. Three rudimentary scientific errors—combined with subversion of the scientific process—have blinded most clinicians and scientists to the connection between acetaminophen and autism.

The issue that we remain less certain of is how long it will take for the reality of the connection between acetaminophen and autism to reach the public consciousness. I am the optimist among our research team, and my guess has already proven wrong. After the president's announcement, I was surprised that I could not get even *one* investigative reporter from a mainstream media outlet to look at the body of available evidence. Fortunately—not being a gambler—I didn't bet much against the guesses of my more pessimistic colleagues.

CHAPTER 11

SUMMARY OF CLINICAL RECOMMENDATIONS

The science connecting acetaminophen and autism is incredibly solid. Not everyone agrees. But it is evident that those who don't agree are misguided—and we can see exactly how and possibly even why they are misguided.

Given the solid science, what should we do in the clinic?

The rules of the empire that govern my actions as a scientist dictate that I cannot give personal medical advice. That is off-limits. On the other hand, as a scientist, it is my job to determine whether medical standards are consistent with available evidence. Science is supposed to be the basis for modern medicine. Basically, I cannot tell a parent or pregnant woman what to do, but it is my job as a scientist to determine what policymaking groups should be telling their physicians, and in turn what physicians should be telling parents and pregnant women.

I cannot tell Nancy Cucumber that she should not use acetaminophen when she delivers her baby. But I can tell anybody that will listen that the best available scientific evidence tells us clearly that *nobody* should *ever* use acetaminophen during labor and delivery.

We published a list of guidelines that caregivers, clinicians, and regulators should adopt concerning the use of acetaminophen during pregnancy, labor and delivery, and early childhood.[347] It's not personal medical advice, but it does say what the medical advice should be. That is the job of scientists, and here are the rules we proposed.

Rule #1: Administration of acetaminophen in a manner that was never intended should be discontinued. This includes treatment of temperatures that do not technically constitute a fever and administration of the drug more frequently and/or at higher doses than recommended.

Rule #2: Administration of acetaminophen under conditions in which evidence demonstrates a lack of effectiveness should be discontinued. This includes the treatment of the pain of circumcision and perhaps the treatment of fevers to prevent febrile seizures.

Rule #3: Administration of acetaminophen under conditions in which no evidence demonstrates long-term benefits of treatment or in which evidence demonstrates a lack of long-term benefits should be discontinued. This includes the treatment of fevers and prophylactic treatments prior to labor and delivery.

Rule #4: Administration of acetaminophen that is no longer recommended by governing medical bodies should be discontinued. This includes the treatment of patients receiving vaccinations[348] and will hopefully include more situations where it is currently administered.

Rule #5: Administration of acetaminophen under conditions where evidence indicates that it is or may be beneficial should not be continued without disclosure of the drug's long-term risks for neurodevelopment. All caregivers, including parents, should be made aware of evidence related to both benefits and risks so that they can make informed decisions.

Those rules should not be overly controversial. Indeed, even the greatest advocate of drug use in babies would approve of Rule #1. In addition, everyone knows that people should be informed about the pros and cons of using any drug, meaning Rule #5 shouldn't be overly controversial.

In talking with clinicians, I find a very, very wide range of opinions about the actual "need" for acetaminophen. When I talk to older physicians, retired or close to retirement, many believe that acetaminophen is not a lifesaving drug. They can remember the time prior to 1980 when acetaminophen was not commonly used. I've also discussed acetaminophen with several "concierge doctors," doctors who specialize in high-end private services who attend functional medicine conferences and integrative medicine conferences. They pride themselves on dealing with the underlying cause of disease and keep up with the latest developments in treatment. Their business model is to stay ahead of the mainstream medical associations, which can take decades to implement changes that the scientific evidence supports. Many of these doctors don't recommend acetaminophen for their patients either. As I explained in chapter 2, the view that acetaminophen induces autism in susceptible children was already evident in 2009 to anyone paying attention to the scientific literature.

And then there are physicians who are convinced that we definitely need acetaminophen. You may recall that, in chapter 4, I mentioned an article by Mary Kekatos published by ABC News on October 14, 2025 ("Debunking 3 Claims about Tylenol After White House Links Drug's Use in Pregnancy to Autism"). That report by Kekatos includes a video from Alok Patel, a pediatrician at Stanford Children's Health. Dr. Patel is a good representative of physicians who believe that acetaminophen is a beneficial drug when given during neurodevelopment. I could have chosen many physicians to bring to center stage here, but Dr. Patel was concise and clear in his delivery. In the introduction, I quoted Paul Klotman at the Baylor College of Medicine. Dr. Patel was just as adamant as Dr. Klotman regarding the safety of acetaminophen, but Dr. Patel gives us a bit more of the rationale for his thinking about medical recommendations for taking acetaminophen. That's what we want to evaluate. Why would a doctor recommend the drug, and does that make sense?

The video featuring Dr. Patel begins by "debunking" the claim that some pregnant women were taking excessive amounts of acetaminophen on social media as a way of defying the Trump administration's warning about acetaminophen being dangerous. The video is a mixture of text, narration, and dialogue from Dr. Patel.

Dr. Patel: We are going to rate this claim as false.

Video text: Doctors who spoke to ABC News said they have not seen any confirmed accounts, or uptick, of women taking acetaminophen in high doses to make a statement.

Dr. Patel: All we've seen is certain pregnant women advocating for the judicious use of Tylenol in certain situations such as a maternal fever or chronic pain. We know that untreated fevers during pregnancy can lead to some long-term developmental outcomes for the developing fetus. And so when having a conversation with a health care professional, it can be deemed that it's totally safe to take acetaminophen in pregnancy, especially because a lot of other pain medications are just not safe to take.

Video text: Trump admin links Tylenol to autism risk. Experts say claim not backed by medical studies.

Video Narrator: They worry that pregnant women will now be afraid to take the drug even when they have a fever, which they say can be more dangerous for the baby.

I understand that this is an ABC News video claiming that the Trump administration is spreading misinformation. I understand that it is not a science article, and I also understand that Dr. Patel is not a scientist. He focuses on public health advocacy, journalism, digital health, and pediatric hospital medicine. Dr. Patel is not an expert in neuroscience, pharmacology, toxicology, biochemistry, statistics, or any other field of science that has anything to do with acetaminophen and autism.

Nevertheless, Dr. Patel is, essentially, representing the medical profession in this video. And he is standing up against "misinformation" from the Trump administration. If somebody is going to challenge the president, who has the FDA and the NIH directors standing behind him as he speaks, they should have evidence. At least *some* evidence.

Although this segment of the video is supposed to be addressing people on social media taking too much acetaminophen, it rapidly digresses into what doctors will recommend based on scientific evidence. So I'll go through the video, piece by piece.

Dr. Patel first states that the president is wrong, that in fact pregnant women on social media have *not* taken too much acetaminophen to spite the president. According to the video, this information was determined by interviewing doctors. As a scientist, I have previously considered how to prove that something never happened. Indeed, we proved that acetaminophen was never proven safe for neurodevelopment.[349] It was a monumental task. To prove that something reported on social media never happened would be almost impossible, if not entirely impossible. Somebody would need to track down the videos and then attempt to find the women who created the videos. If the women or even the videos were not to be found, one could only say that the rumors are *unconfirmed*, not that they are *false*. It may be possible to determine that one video or one story was fake, but *all* of them?

Regardless of whether every person could be tracked down, interviewing doctors is not how a reasonable social scientist would set about trying to verify that events reported on social media did or did not happen. During my scientific career, I conducted a number of "socio-medical" studies aimed at determining what doctors observe in their patients.[350] Even if a doctor knew a patient who took too much acetaminophen while pregnant, a law, called HIPAA, prevents doctors in the United States from talking about specific patients. This is an exceedingly well enforced law. No doctor would ever violate that law for the sake of an interview. As a scientist who has worked on research involving human subjects, I have the same training pertaining to that law that physicians have. That law is hammered into our subconscious. Any serious social scientist trying to track down a patient who took too much acetaminophen during pregnancy to spite the Trump administration would have a difficult time if they approached doctors first. They would know not to bother approaching doctors.

To get the ball rolling, Dr. Patel, without having done any legitimate research to back it up as far as we can tell, calls out the president for making a false statement. That first conclusion by Dr. Patel is unfounded, but if anyone believes him, then it undermines the president's credibility for any additional claims he may have made.

The next thing Dr. Patel does is to claim that women want to use acetaminophen in "judicious" amounts. I would assert that women would want to know what the advantages and disadvantages of taking acetaminophen are. That is my experience. Pregnant women are rational humans and want to be informed about the evidence. My experience is that they *don't* want to take a drug if the evidence says it might cause permanent harm to their child. Dr. Patel's second claim is nonsensical.

Next, Dr. Patel insinuates that taking acetaminophen during pregnancy is "totally safe." He doesn't actually say it's totally safe, but that your physician might *deem* it totally safe. Dr. Patel is technically saying that some doctors are unaware of the risks—and that we already all know. But many people could take the message to mean that acetaminophen *is* totally safe. The message is technically accurate, but entirely misleading.

Dr. Patel then proceeds to state an overt logical fallacy. He claims that because *other* medications are deemed unsafe, then acetaminophen is especially likely to be considered safe. A very effective drug once used to treat morning sickness is no longer used during pregnancy, despite the fact that there is no suitable replacement. That drug, thalidomide, causes birth defects, and the fact that it was ever used is considered a tragedy.[351] Just because no other drug is available does not mean that acetaminophen is safe.

The next statement in Dr. Patel's video says that experts say acetaminophen is not linked to autism. This entire book is about the tragic error of that statement; thus, it will not be reviewed here. Here we want to focus on medical recommendations.

The final statement in this segment of the video claims that a fever *without* acetaminophen is more dangerous for a baby than a fever *with* acetaminophen. This is not a wildlife documentary where somebody is required to observe the feverish baby or pregnant mother-to-be without doing anything. Underlying conditions that might be life-threatening should be treated, and the individual can be made comfortable with a wide range of interventions. Here we are considering the risks of a fever *without acetaminophen*, not the risks of a fever with no care whatsoever.

Nothing in the scientific literature supports Dr. Patel's claim that acetaminophen makes a fever safer. My research group has literally checked every study in the entire scientific literature claiming that acetaminophen is safe for pediatric use.[352] None of those studies identified long-term *benefits* from acetaminophen use. If acetaminophen had actually helped to improve long-term outcomes, it would have been a highly touted medical discovery.

More importantly, we know that fevers are a protective response from our own immune system,[353] that blocking fevers suppresses the immune system,[354] and that fevers are not generally dangerous.[355] The problem is that the things that cause fevers, primarily infections, can be dangerous.

The important point here is not to confuse association with causation. What causes the fever may be dangerous, but the fever itself is part of the body's efforts to defend itself.

After "debunking" the first claim made by the president, Dr. Patel goes on to claim that increases in the prevalence of autism can be explained by (a) changing definitions of autism, (b) more awareness, and (c) more screening. That false claim was addressed in chapter 4 and will not be covered here. Next Dr. Patel refers to the Drexel study and claims there is no evidence that acetaminophen causes autism. That false claim was addressed in chapter 3 and will not be covered here. In this chapter, we are sticking with medical recommendations.

Finally, Dr. Patel debunks the president's claim that "You should not give acetaminophen to young children, to babies."

> **Dr. Patel:** I'm also going to call this claim false, with an asterisk. We'd like to reduce the amount of medications we're potentially giving to any newborns or babies, but there are situations in which we do recommend that babies, yes, even babies, get acetaminophen. There are situations when babies have fevers. They're in pain, they're cranky, they're irritable from a whole host of things. And giving them acetaminophen might mean the difference between them sleeping and not sleeping or them drinking breast milk and eating or not having anything at all. And guess what? A calm baby who is sleeping and hydrated is a much safer place to be than a baby who's left untreated and uncomfortable.

Here Dr. Patel is saying we should reduce the use of all medications, and at the same time saying we should use acetaminophen for literally "a whole host of things." More importantly, Dr. Patel is saying that a baby who has acetaminophen permeating his or her tiny brain is in "a much safer place" than an *uncomfortable* baby. Here Dr. Patel is breaking **Rule # 1**. Being uncomfortable or "cranky" was never an accepted reason to give acetaminophen. It's just not on the list.

Dr. Patel apparently doesn't know that the baby converts acetaminophen into a toxic metabolite, that acetaminophen is difficult for newborn babies to safely metabolize, and it is extremely toxic to brain development in laboratory animals. We can tell from the rest of the video that he also doesn't know acetaminophen is associated with autism in numerous ways through time and place—assuming we don't adjust away those associations using invalid statistical manipulations.

As more parents become aware of the dangers of acetaminophen for neurodevelopment, the magnitude of this medical error will be realized. I am grateful to the parents of four children, three boys and one girl, for describing their experience and allowing me to record it here. Their three oldest children, two boys and the girl, were all exposed to acetaminophen regularly during pregnancy, each child receiving an estimated fifteen to twenty days of total exposure prior to birth to treat pain and fevers. The same three children also received acetaminophen prior to each vaccination. In addition, one of the boys was exposed to acetaminophen at the time of circumcision, at about three days of age, and treated with acetaminophen at home after the circumcision. The other boy was exposed to multiple doses of acetaminophen for several days at one month of age due to a severe ear infection. All three older children displayed behaviors associated with autism by eighteen months of age. Behaviors were manifested differently in each child and included heightened sensitivity to new or strong stimuli, repetitive motions, difficulty with emotional regulation, and developmental delays (speech and toilet training). Eventually, the two oldest children (one boy and one girl) were formally diagnosed with autism. The third child, now a teenager, has a very literal mindset, struggling with figurative language, including similes and metaphors. He is presumed to have autism based on the observations of his physicians and parents, although he has never been formally tested.

Before the fourth child was conceived, the parents became aware of a possible connection between acetaminophen and autism. This child was given a single dose of acetaminophen at the age of one year to treat a persistent fever, but he was otherwise never exposed to acetaminophen. The mother described the avoidance of acetaminophen during pregnancy as "extremely difficult," especially given her age; her fourth child was born more than a decade after her third, and, she explained, the aches and pains of pregnancy can increase with age. The boy, now two years of age, has no behaviors characteristic of autism. The parents point out that they are now very aware of early indicators of autism, and have paid very close attention to their youngest child's development, which appears to be neurotypical and much different than that of his older siblings.

When considering what the standard of medical care should look like, physicians should keep in mind that the time immediately around birth, starting at the instant the umbilical cord is clamped and extending for about two months or so, is apparently the most critical. With rates of autism now at 3 percent, any use of acetaminophen during that time frame should be only for saving the life of the mother or child. Yet, no such life-saving effects of acetaminophen have been documented. Even spiking a fever during labor and delivery, a condition that can cause an elevated heart rate in the fetus on rare occasions, can be alleviated using delivery by C-section.

We should also consider the fact that about half of the regressions into autism occur by eighteen months of age. That timing agrees well with the pharmacological data we discussed in chapter 5, which indicates that babies and toddlers don't gain their full ability to process acetaminophen until about twenty-four months. At the same time, however, some children are more susceptible for reasons that we understand—and will remain at risk until they reach about six years of age.

With that in mind, it is very, very difficult to imagine any instance in which acetaminophen should be used between birth and age six. The benefits would have to be tremendous, and no such benefits have been reported. Thus, I would conclude that the president's advice makes sense, and we should avoid acetaminophen exposure during that time.

Reasonable guidelines for acetaminophen use during pregnancy may be more difficult to establish because we have less certainty regarding risks during pregnancy. Based on the study from Drexel[356] —ignoring the fact that they mangled the interpretation of their beautiful data—it seems likely that use of acetaminophen during pregnancy for chronic pain management carries some risk. It's much less clear that occasional use to treat fever during pregnancy would cause a significant risk. A significant association was found, but the association was small and could be explained by other factors. Perhaps, for example, women who occasionally use acetaminophen during pregnancy may be more likely to give it to their baby. The bottom line is we just don't know whether the mother's liver can provide complete protection from acetaminophen when used occasionally during pregnancy.

Scientists could evaluate the safety of acetaminophen during pregnancy in a prospective study. The study team would need to monitor all acetaminophen use up until the age of about six years, not just during pregnancy. That's never been done. Because we know the drug is potentially dangerous, the scientists cannot ethically do the experiment blinded, and they can't randomize people to different treatment groups. These are normal components of most drug studies. But the NIH could probably conduct an incisive study if they decided that acetaminophen use during pregnancy was worth the vast amount of time and energy it would take. Then we could answer questions more definitively about risks during pregnancy. For now, it's a mystery. And parents-to-be should know that it's a mystery. Bluffing in order to alleviate fear and worry is a bad idea. If we don't know, we don't know.

I've talked to several concierge doctors about how they recommend treating fever and pain. Some of them immediately, without thinking, list non-pharmaceutical options they believe are effective, even for ear infections. When I ask doctors in major medical centers the same question, they tell me that nothing has been tested and approved. They aren't wrong. Testing alternatives is something that, again, the NIH could be doing. That would be a really good use of their resources and would help support lots of clinician scientists.

I'm frequently asked if taking metabolic supplements could alleviate the neurodevelopmental risks associated with taking acetaminophen. Based on my understanding of the metabolic pathways the body uses to safely process acetaminophen, I would guess (hypothesize) that it might be possible to enhance those pathways. However, we would need to grapple with some ethical questions before that research could be done. Since we know acetaminophen is a developmental neurotoxin, how could we intentionally expose any baby to the drug, even if we thought our "antidote" would block toxicity? I don't know if we want to know the answer to that question enough to do the experiment. I imagine that we first need to decide whether treating pain and fever with acetaminophen is worth the risks of using it even in an experimental study. Based on current standards, we need to try any metabolic support for acetaminophen use in laboratory animals before trying it in humans.

There is much we don't know. But we know more than enough to guide dramatic changes in current medical practice. I cannot give personal medical advice, but I can point out information that needs to be known by anyone considering the use of acetaminophen during neuro-development before the age of six years:

- Use of acetaminophen from labor and delivery until age six should be "contraindicated" because the known risks outweigh the known benefits.
- No scientific evidence demonstrates that the benefits outweigh the risks of acetaminophen use for any reason between conception and age six. When considering such studies, medical treatment of the control (no acetaminophen) group is critical. Unfortunately, many potential alternatives to acetaminophen involving non-pharmaceutical interventions have not been thoroughly tested.
- An overwhelming body of evidence tells us that the harm from current acetaminophen use during brain development vastly outweighs any benefits that we might receive from the drug. And by vast, I really mean vast—really vast indeed.
- A course correction in clinical practice is long overdue.

APPENDIX A

Handling by the *Journal of Xenobiotics* (*JoX*) of the following manuscript prior to its publication in the *Journal of the Academy of Public Health*:

Parker, W., Corrigan, P.T., Anderson, R., Jones, J. P., III, Konsoula, Z., Williamson, L., Bollinger, R.R. & (2025) Evidence that acetaminophen triggers autism in susceptible individuals has been ignored and mishandled for more than a decade. *Journal of the Academy of Public Health.*

The unedited correspondence between William Parker and the editors, including the reviews, the letter of protest, and the final rejection of the manuscript are included in their entirety. Contact information has been deleted.

Initial rejection from *JoX*, including reviewer comments.
On Sept. 1, 2025, at 4:08 a.m., JoX Editorial Office wrote:
Dear Professor Parker,

Your manuscript has been rejected for publication in JoX:

Manuscript ID: jox-3823193
Type of manuscript: Review
Title: Evidence that acetaminophen triggers autism in susceptible individuals has been ignored and mishandled for more than a decade
Received: 31 Jul 2025

This decision was based on the comments of external experts who carefully reviewed your manuscript. For your consideration, the review reports can be found below.

Should you wish to submit your manuscript elsewhere for publication, you are welcome to incorporate the reviewers' suggestions. If your manuscript requires improvement to the language and/or figures, you may consider MDPI Author Services: https://www.mdpi.com/authors/english

We thank you for considering JoX to publish your work and wish you every success in the future.

If you have any questions, please do not hesitate to contact us.

Kind regards,
JoX Editorial Office

Reviewer #1
"Evidence That Acetaminophen Triggers Autism in Susceptible Individuals Has Been Ignored and Mishandled for More than a Decade" is a review of the literature on a possible association between acetaminophen use in the first years of life and autism risk. The review is comprehensive in identifying all the literature on an association between acetaminophen use in the first years of life and autism risk, then summarizing the clinical recommendation in the conclusion of the paper. The review is correct and agrees with other recent reviews (see, for example, PMID 40804730) that there is a documented association between acetaminophen exposure during pregnancy/early postnatal life and diagnoses of neurodevelopmental disorders such as autism and ADHD. The review is also correct that, despite what is now substantial scientific evidence of an association, there has been an appalling lack of guidance from the relevant US government health agencies to limit or avoid acetaminophen exposure.

However, I recommend rejection of this paper for publication because its a priori conclusions are outrageous, overstated, illogical, and unnecessarily inflammatory. The first line of the Abstract says: "Overwhelming evidence shows that exposure of susceptible babies and children to acetaminophen (paracetamol) triggers many if not most cases of autism spectrum disorder." That is simply illogical and untrue. A 2022 paper from the same senior author (PMID 35175416) reported in the "What is Known" section that "Paracetamol (acetaminophen) is widely thought by pediatricians and parents to be safe when used as directed in the pediatric population, and is the most widely used drug in that population, with more than 90% of children exposed to the drug in some reports." As cited in that 2022 paper, as many as 95% of children are exposed to acetaminophen by age 9 months (PMID 22253317). Nearly every child is exposed to acetaminophen. Thus, saying that acetaminophen exposure causes autism in susceptible children is the logical equivalent to saying that breathing air causes autism in susceptible children. Nearly every single child is exposed to acetaminophen, yet only 3% of children have autism. Why are the other 95+% not susceptible to acetaminophen exposure? Nobody can be confident that an exposure causes a phenotype unless 100% of those exposed exhibit the phenotype. The authors try to argue that "susceptibility is imparted by inflammation and oxidative stress, which can be caused by a wide range of genetic, epigenetic, and environmental factors" (Introduction, second sentence, lines 38-40). That may be true. But until the genetic, epigenetic, and environmental factors are identified that CAUSE the susceptibility, it is not rational to make the argument that acetaminophen exposure is causal only in those that are susceptible, an unidentifiable subset of the population. Thus, the claim in the title that "Acetaminophen Triggers Autism" is false until the underlying reason for the susceptibility is discovered.

The review also rages against straw enemies. From the Abstract: "pervasive were mishandlings" and "erroneous criticisms." These are accusations that are unwarranted.

In summary, this is a review of a possible association between acetaminophen exposure and autism risk. While there is some evidence of an association, the review is illogical at its inception. Acetaminophen use is so pervasive—90 to 95% of infants are exposed—that the claim that it "triggers many if not most cases of autism" is simply not a rational conclusion. The same literature can be used to make a strong argument that the evidence of an association is sufficient to warrant government action to limit or prevent acetaminophen exposure during the early years of life. The tone of the present review does not make a strong argument.

Reviewer #2

Well written systematic review of the literature concerning potential connections between acetaminophen use with increased risk of autism.

Comments

1. The manuscript is quite repetitive, especially in the Discussion section.
2. The authors should define what they mean by "susceptible" children—presence of oxidative stress is not sufficient definition.
3. The authors should define what they mean by oxidative stress and give examples.
4. Oxidative stress is usually associated with inflammation. The authors should discuss if any other allergic or inflammatory conditions constitute "susceptibility"—see example below

Theoharides TC, Tsilioni I, Patel AB, Doyle R. Atopic diseases and inflammation of the brain in the pathogenesis of autism spectrum disorders. *Transl Psychiatry*. 2016;6(6):e844. Published 2016 Jun 28. doi:10.1038/tp.2016.77

5. The mechanism of acetaminophen neurotoxicity should be presented and discussed—see example below

Bazan HA, Giles BL, Bhattacharjee S, Edwards S, Bazan NG. A non-toxic analgesic elicits cell-specific genomic and epigenomic modulation by targeting the PAG brain region. *Neurobiol Pain*. 2025;18:100192. Published 2025 Jul 20. doi:10.1016/j.ynpai.2025.100192

6. The screening for P450 enzyme polymorphisms that may define slow acetaminophen metabolizers as "susceptible" should be discussed.
7. The use of N-acetylcysteine (NAC) for children suspected of given too much acetaminophen should be recommended. See example below

Chen F, Lu SZ, Choha H, Lau A. The Role of Fomepizole in Acetaminophen-related Poisoning: A Narrative Review. *Clin Exp Emerg Med*. Published online August 13, 2025. doi:10.15441/ceem.25.059

Response to Journal from senior author (WP)
From: William Parker
Sent: Monday, September 1, 2025 8:31 a.m.
To: William Parker; Veralyn Kong
Cc: JoX Editorial Office; Zara Liu
Subject: Re: [JoX] Manuscript ID: jox-3823193 - Declined for Publication

Dear Editor Kong,

Thank you very much for processing our manuscript.

We are grateful to the second reviewer, whose comments were helpful.

The first reviewer is very convinced that the conclusions underlying our work cannot possibly be accurate, and that in fact are "outrageous" and "illogical," as described by the reviewer.

Although explained in the Introduction of the manuscript, I would like to remind the editor that the evidence for the conclusions underlying our work have been peer reviewed, and are incontrovertible. For the editor's convenience, we have attached a brief summary of that evidence.

Since the reviewer repeatedly states that the science underling our work is incorrect but does not address any published evidence, the reviewer's view is, by definition, a "fallacy of invincible ignorance." The review can also accurately be described as an argument from personal incredulity.

We invite the editor to reconsider the manuscript, seeking a reviewer who is familiar with the evidence underlying the work described in the manuscript.

We do understand if the editorial decision is final. If that is the case, please let us know.

Thank you again for your time and efforts in processing our manuscript.

Sincerely,
William Parker, Ph.D.

Final rejection from the Editor

On Sept. 17, 2025, at 5:27 a.m., Veralyn Kong wrote:

Dear Prof. Parker,

I hope this letter finds you well.

A few days ago, we submitted your appeal to our Academic Editor for a re-evaluation. We regret to inform you that we can not further proceed your manuscript after the evaluation. We will reject your manuscript by our system soon. We wish you every success if you choose to pursue publication elsewhere.

Please feel free to contact me if you have any questions.

Kind regards,
Ms. Veralyn Kong
Assistant Editor

EVIDENCE SUMMARY TABLES

Table 1. Six lines of *pharmacologic evidence* connecting acetaminophen and autism. APAP = acetaminophen, ASD = autism spectrum disorder.

Pharmacologic evidence / references	Background / additional information
1. Mechanisms of APAP-mediated injury are plausible. For review, see Jones et al.[357] and Figure 2.1.	The first study showing that children with ASD are deficient in a metabolic pathway necessary to safely detoxify APAP in babies (sulfation) is now more than a quarter of a century old,[358] and was subsequently corroborated.[359] Another pathway necessary to safely detoxify APAP is deficient in newborns[360] and may also be especially deficient in children with autism.[361] One enzyme (CyP450 2E1) which produces the toxic metabolite of APAP (NAPQI) is expressed in the human brain from before birth[362] and can be different in mothers who have children with ASD.[363] In addition, differences in another enzyme (CyP450 1A2) that produces the same toxic metabolite of APAP is associated with ASD.[364]
2. Increasing evidence suggests that the gut/brain axis may play a role in many cases of ASD (reviewed recently),[365] and APAP is known to adversely affect gut function in laboratory animal models[366] and possibly in humans.[367]	Aberrant gut function leading to oxidative stress and inflammation is among many factors that would predispose individuals to adverse reactions to APAP leading to ASD,[368] and gut microbial metabolites with structures related to APAP serve as excellent biomarkers for ASD.[369] However, it remains unknown whether aberrant gut function can be induced by APAP at the time of ASD induction and play a role in that induction.
3. Genetic, epigenetic, and environmental factors associated with an increased risk of ASD have an adverse effect on the body's ability to safely metabolize APAP.[370] See Figure 2.2.	The wide array of factors associated with ASD have led to the hypothesis that many things can come together to cause ASD, but ASD is characterized by impairment of social function and by particular behaviors, suggesting specificity in the cause of the condition.

4. Cystic fibrosis is associated with unusually efficient (effective) metabolism of APAP,[371] and the prevalence of ASD is apparently very low in patients with cystic fibrosis.[372]	The mental health of patients with cystic fibrosis has been characterized extensively, but no association between ASD and cystic fibrosis has been reported.
5. Calcium signaling is known to be important in ASD,[373] and APAP adversely affects calcium signaling.[374]	Calcium signaling dysfunction can be used as a biomarker for ASD,[375] although it remains unknown whether APAP plays any role in inducing that dysfunction.
6. APAP binds directly to arachidonic acid[376] and affects arachidonic acid metabolism.[377] Alterations of arachidonic acid[378] and enzymes involved in arachidonic acid metabolism[379] are associated with ASD.	It is unknown what role arachidonic acid plays in ASD, but arachidonic acid plays a role in both the analgesic and antipyretic properties of APAP, and its metabolism is associated with ASD.

Table 2. Twelve associations connecting acetaminophen and autism through either *time, place,* or *human behavior*. The numbering is continued from Table 2.1. APAP = acetaminophen, ASD = autism spectrum disorder.

Associations involving time, place, or human behavior and references	Background / additional information
7. APAP use during early childhood is associated with a twenty-fold greater risk of regressive ASD.[380]	This case-controlled study, now more than 16 years old, has been widely criticized, but careful analysis does not reveal any credible objections.[381]
8. APAP use with mild adverse reactions to a vaccine, but not mild adverse reactions to a vaccine alone, is associated with ASD.[382]	This study, the same as the study providing line of evidence #7, was the first study to separate the impact of vaccines from APAP on neurodevelopment, and the first to implicate APAP in the development of ASD.
9. Higher levels of APAP in cord blood are associated with ASD.[383] See chapter 5.	For the analysis, the authors divided the women into three groups based on cord blood APAP levels. The third with the highest levels had 3.6 times more likelihood of having a child with ASD than the third with the lowest levels of APAP.
10. Use of APAP during pregnancy has been associated with adverse long-term effects on the mental health of offspring in numerous studies.[384] See chapter 3.	This line of evidence has received more attention than any other line of evidence, to the point of being the only line of evidence considered by many investigators. However, the numerous studies underpinning this line of evidence are hampered by several factors which can cause errors in estimation of the association between APAP and ASD.[385] A recent study found a dramatic association (odds ratio (OR) for ASD with APAP use = 1.8),[386] but incorrectly and completely canceled out that association using an error in the assumptions underlying the statistical analysis.[387]

11. Analysis of the Danish National Birth Cohort (DNBC) revealed a 30 percent increased risk for ASD associated with postnatal APAP exposure,[388] despite the fact that the use of APAP appears to be dramatically underreported in the DNBC.[389]	The study authors averaged the results from the DNBC with assessments of autism-like symptoms (not ASD) from smaller data sets, and reported no association between postpartum APAP use and those symptoms (not ASD) in the abstract of the paper.[390] This issue has been addressed in detail by us in the literature,[391] but unfortunately may still result in confusion.[392] In addition, the study[393] employed invalid statistical adjustments that underestimated the association between APAP and ASD.[394] See text for additional discussion.
12. The incidence of ASD began to increase in the early 1980s, coinciding with the increase in APAP use after aspirin was associated with Reye's syndrome.[395] See chapter 4.	Temporal associations do not prove causality but are necessary for causality to exist. Alternative explanations for the rise in prevalence of ASD face several insurmountable problems, previously reviewed.[396]
13. The incidence of ASD has steadily increased[397] as marketing of acetaminophen was aggressively pursued[398] and perhaps other factors such as mandated use of APAP with the MenB vaccine have led to increased APAP exposure early in life.	Temporal associations do not prove causality but are necessary for causality to exist. Alternative explanations for the rise in prevalence of ASD face several insurmountable problems, previously reviewed.[399] One possible explanation for the persistence of unrealistic alternative explanations may be that many investigators are unaware of a satisfactory explanation consistent with available evidence.
14. The ratio of regressive to infantile ASD rose at the same time as pediatric APAP use rose[400] after aspirin was associated with Reye's syndrome.[401]	This observation, made in 2000, would suggest that something was introduced into the environment that could induce ASD after months or even years of neurodevelopment. This factor was incorrectly suspected to be vaccines at that time, an issue that was decisively addressed by Stephen Schultz eight years later (see line of evidence #3).
15. Circumcision of males is associated with a twofold increase in the risk for early-onset (infantile) ASD.[402]	Studies in Denmark[403] and the United States[404] have identified social problems associated with circumcision. Circumcision is often performed using APAP as an analgesic, despite the fact that such use is of highly questionable effectiveness.[405] Parents are sometimes advised to use acetaminophen after the procedure, which can cause discomfort for twenty-four hours or longer.
16. The popularity of APAP use and the prevalence of ASD were substantially higher in Denmark than in Finland in the mid-2000s.[406]	Geographic-dependent associations do not prove causality but do contribute to the total body of evidence. Particularly in the absence of alternative explanations, these associations can be compelling.
17. Ultra-Orthodox Jews[407] in Israel have a reported prevalence of ASD less than half of that of reform Jews. Traditional circumcision practices employed by Ultra-Orthodox Jews do not utilize APAP. Further, Orthodox Jews may use less pain medication during childbirth than others.[408]	Circumcision is often performed using APAP as an analgesic, despite the fact that such use is of highly questionable effectiveness.[409] Almost all Israeli Jews are circumcised.[410] Further, decreased use of pain medications during labor and delivery, which may be associated with Orthodox Judaism,[411] would likely lower exposure to APAP at the time of life when individuals are the most sensitive to APAP-mediated neurodevelopmental injury.[412]

18. Studies in several countries with chronic shortages of medication found dramatically lower-than-expected levels of ASD relative to other developmental issues, including Down syndrome. Reviewed by Jones et al.[413] See chapter 4.	In addition to evidence presented in the previous review of this issue,[414] apparently low levels of ASD in Cuba have been identified, where 241 cases of ASD in the entire nation (1 in 25,000 children) were identified in a 2016 report.[415] APAP is available in Cuba by prescription only,[416] and multiple travel advisors cite APAP in particular as being in short supply in Cuba.[417] In addition, individuals with pervasive developmental disorders, about 50 percent of which typically have ASD, were found among second generation Israeli immigrants from Ethiopia, but not among first generation Israeli immigrants from Ethiopia ($p = 0.0022$).[418]

Table 3. Four lines of evidence from laboratory animal studies indicating that acetaminophen is toxic to the developing brain. In total, about twenty different studies support the four lines of evidence. The numbering is continued from Table 2.2. APAP = acetaminophen, ASD = autism spectrum disorder.

Laboratory animal studies and references	Background / additional information
19. Numerous studies in laboratory animals from multiple laboratories indicate that early life exposure to APAP causes long-term changes in brain function.[419]	After adjusting for weight, the amount of APAP that causes profound changes in laboratory animals in some studies is very close to[420] or even less than[421] the amount administered to human babies and children. Thus, APAP could never be used in babies or children if current guidelines for drug safety were applied.
20. Early life exposure to APAP has a greater long-term impact on male laboratory animals than female laboratory animals.[422] ASD is more common in males than in females.	The reason or reasons why males are more susceptible to APAP-mediated injury has been considered in some detail, and several plausible mechanisms have been proposed.[423]
21. Prenatal exposure of laboratory rats to APAP causes problems with the processing of sound[424] and with olfactory function.[425] Impairment of olfactory function[426] and some degree of auditory dysfunction (reviewed by Graeca and Kulesza)[427] are associated with ASD.	It is unknown whether these effects of APAP on laboratory animals are related to altered processing of smell and hearing in some individuals with ASD.
22. APAP causes apoptosis-mediated death of cortical neurons in laboratory rats,[428] and cortical neurons may be involved in the pathology of ASD.[429]	Increased levels of biomarkers for apoptosis,[430] a type of brain cell death, and impaired autophagy,[431] the means by which the body cleans up dead cells, are associated with ASD. Autophagy is also necessary for clearing damaged organelles such as mitochondria,[432] which are created by aberrant metabolism of APAP.[433]

Table 4. Eight lines of miscellaneous evidence connecting acetaminophen and autism. The numbering is continued from Table 2.3. APAP = acetaminophen, ASD = autism spectrum disorder.

Miscellaneous evidence and references	Background / additional information
23. APAP was never demonstrated to be safe for neurodevelopment.[434] Over two hundred papers in the medical literature claim that APAP is safe for babies and/or children when used as directed, but all studies were based on the false assumption that adverse reactions in babies would involve easily measured liver injury, the same as in adults.[435]	Like APAP, opioids have also never been shown to be safe for neurodevelopment.[436] However, unlike APAP, opioids are not generally *assumed* to be safe for neurodevelopment when used as directed. Further, one study probing the safety of prenatal opioid exposure found reductions in communication skills in children associated with prenatal APAP exposure, but not with prenatal opioid exposure.[437]
24. APAP inhibits neuronal cell growth in tissue culture experiments, altering "arborization," the process by which neurons branch out to make connections with other neurons.[438] APAP[439] and a metabolite of APAP[440] also cause death of brain cells in culture.	Adverse effects of APAP on neuronal cell growth in culture (*in vitro*) are dose dependent and observable at concentrations near those achieved in clinical therapy.[441] These effects *in vitro* would discourage use of APAP in humans if current guidelines for drug safety were applied.
25. In adults, APAP temporarily blunts social trust[442] and awareness,[443] emotional responses to external stimuli,[444] and the ability to identify errors.[445]	Although the mechanisms are unknown, these studies show that APAP affects aspects of mental function that are impaired in individuals with ASD.
26. APAP is not used in domestic cats because they lack a robust glucuronidation-dependent capacity for metabolism,[446] making them susceptible to APAP-mediated toxicity. Human newborns also lack a robust glucuronidation-dependent pathway.[447]	Based on liver function in human babies and children, APAP was incorrectly determined to be safe for pediatric use in the 1960s and 1970s (see line of evidence # 4), before this evidence from veterinary science became available in the 1980s. One study in laboratory animals in the 1980s showed that even lethal doses of APAP do not cause liver failure in neonates,[448] but the first study showing APAP-mediated neurodevelopmental brain injury in laboratory animals was not published until 2013.[449]
27. Surveys show that up to 50 percent of parents who have a child with ASD believe that their children's ASD was induced by one or more vaccines.[450]	Although this belief has been widely attributed to a 1998 report describing 12 patients,[451] the title of that report is not intelligible to individuals outside of the medical profession, and medical papers have seldom affected public opinion. A more likely explanation involves the induction of ASD by APAP use concurrent with vaccination, as suggested by Schultz.[452]

28. The "missing heritability" paradox of ASD suggests that epigenetic factors or very early exposure to environmental factors might influence the onset of ASD.[453]	The role of APAP in the induction of ASD nicely resolves the missing heritability paradox connected with ASD, in which sibling studies indicate a high contribution of genetics, but genome wide studies fail to identify the genes involved.[454] The observation that abuse of a mother when she was a child is associated with ASD in the offspring[455] is one example of evidence that supports this view.
29. ASD and fetal alcohol spectrum disorder (FASD) are similar in many regards. Reviewed by Jones et al.[456] A spectrum disorder can also be triggered by the drug valproate.[457]	These observations demonstrate that a complex spectrum disorder (FASD) sharing many similarities with ASD can (a) be induced by a single chemical and (b) be influenced by a variety of genetic and environmental factors.
30. Common alternative explanations are not consistent with known observations and/or require elaborate/complex scenarios to be true.	Some alternative explanations depend on the view that the incidence of ASD has not increased dramatically over time, a view that is contradicted by numerous lines of evidence showing that ASD is an effect of industrialized culture. Evidence that rules out alternative explanations for the etiology of ASD have been described previously.[458]

FIRST ROUND OF REVIEWS AND REBUTTALS FROM *PEDIATRICS*

The unedited correspondence from William Parker to the editor, including the letter to the editor, the reviewer comments (bold font), and the responses to the reviewer comments, is included in its entirety.

Dear Professor Kemper,

On behalf of all authors, I offer our sincere thanks for allowing us to submit the perspective and for obtaining two very intriguing reviews. While both reviewers are adamant that our conclusions are wrong, neither review provides any evidence suggesting that we could possibly be wrong, and both reviewers consistently misrepresent the basis for our conclusions. The second reviewer attempts to present an argument based on evidence that our conclusions are wrong, but the argument contains a critical error that has already been pointed out in the literature and ignores more than 20 additional lines of evidence. Further, based on specific comments, it seems apparent that the second reviewer in particular is entirely unfamiliar with the three recent reviews which provide the basis for the perspective. Given the importance of this topic, we would hope that reviewers familiar with the topic could be found.

I provide a point-by-point rebuttal to the reviewer's comments below.

Our team includes expertise in toxicology, pharmacology, neuroscience, immunology, and biochemistry with a century of combined experience in science. I assure you that our conclusions are based on overwhelming evidence with a lack of any evidence to the contrary. With this in mind, I believe that practice will eventually change, and I hope that *Pediatrics* will be at the forefront of this impending historical change. We see this perspective as an opportunity to course-correct practice in the field.

We request that, based on the fact-based rebuttal below, you will reconsider your decision to reject the perspective. We would also request that any further reviews on this vitally important topic be subject to fact-checking.

We thank you again for giving us the opportunity to submit our work to *Pediatrics*.

Sincerely,
William

William Parker, PhD

Visiting Scholar, Department of Psychology and Neuroscience, University of North Carolina, Chapel Hill
CEO, WPLab, Inc., a 501(c)3 nonprofit conducting research and education on the cause of chronic inflammatory disease in high income areas of the world
Associate Professor, Duke University, retired

Reviewer comments are in bold. Responses are in plain font. Numbers were added for clarity and cross-referencing. Citations are limited to those used in the perspective and thus in some cases refer to a review of the original study rather than the original study.

Reviewer 1 Comments for the Author:
1.1. In this Pediatric Perspective, the authors claim that acetaminophen exposure accounts for the majority of cases of ASD and provides a more compelling explanation than "the multifactorial model."
This characterization of our perspective is somewhat misleading. We are in fact stating that incorporating acetaminophen into, not in lieu of, the "multifactorial model" is the best explanation for all available observations.

1.2. Their "evidence" to support this assertion primarily comes from several papers from their own group that provide circumstantial claims about acetaminophen as a causal agent in the pathogenesis of autism.

This characterization of our evidence is grossly misleading in two regards. First, the large majority of evidence does not actually come from our group. Of the 22 lines of evidence, only two lines include any of our group's work, and only one of those lines of evidence is derived exclusively from our group: the formal proof that acetaminophen was never shown to be safe for pediatric use [1]. Our studies in laboratory animal models are supported by several other groups, and thus the line of evidence derived from laboratory animal studies such as ours do not depend entirely on our group's results, as pointed out in all three of our recent reviews of the topic [2–4]. Second, many of the 22 lines of evidence cannot be classified as circumstantial. For example, none of the following lines of evidence are circumstantial:

(a)   two lines of evidence from direct associations between acetaminophen exposure and ASD

(b)   pharmacological/toxicology studies of the mechanism of acetaminophen-induced injury

(c)   the temporary impact of acetaminophen on social awareness in adults

(d)   three lines of evidence based on studies testing the neurodevelopmental impact of acetaminophen in laboratory animals

(e)   proof that acetaminophen was never shown to be safe for neurodevelopment

We understand that some lines of evidence, including associations of ASD with human behavior (i.e., circumcision) and temporal associations between ASD and acetaminophen use are circumstantial. At the same time, viable alternative explanations for the circumstantial evidence are lacking, invoking the application of Occam's razor, as described in the perspective. Further, circumstantial evidence can be convincing when multiple, independent lines of evidence are available, and when supported by non-circumstantial lines of evidence.

1.3. The certainty with which they make this claim is not supported by existing scientific evidence and ignores the complexity of what is currently known about the epidemiology, etiology and clinical course of ASD.

This comment is verifiably false. As pointed out in our perspective, inclusion of acetaminophen in the model not only incorporates the complexity of the currently accepted model, but also explains current observations regarding ASD better than the currently accepted model alone. Indeed, the perspective points out that "A tally published in 2022 [4] found six unknown factors that must be invoked and eight largely independent observations that must be attributed to coincidence if acetaminophen is not a neurodevelopmental toxin."

For example, when considering the "complexity of what is currently known," how is it possible to explain a 2-fold increase in infantile ASD associated with a simple medical procedure (circumcision) performed in the first days of life, and the 3.6-fold increased prevalence of ASD associated with the upper tertile of individuals based on acetaminophen-associated cord blood metabolites? These are only 2 examples of 22 examples we have identified, with no indications of any weaknesses in the model based on what is currently known.

1.4. The authors' call to action makes some reasonable suggestions that are in fact already part of standard practice in many pediatric practices- e.g., assuring parents that fevers are not dangerous, and not routinely giving acetaminophen for pain at the time of circumcision.

This comment misrepresents our call to action in three regards. First, we call for discontinued use during the peripartum period, which, as pointed out by the second reviewer (points 2.3 and 2.11), is novel. Further, for reasons not provided, this reviewer does not accept the conclusions we have reached regarding the role of acetaminophen in the induction of ASD. For that reason, it is doubtful that this reviewer supports our recommendations for informing caregivers of the role of acetaminophen in the induction of ASD. Finally, while we do recognize that some of our suggestions are already considered standard of practice for some clinics, all available evidence points towards continued use of acetaminophen under conditions in which it is either not shown to be effective or in which it is not needed. Thus, the education piece of our recommendations is vitally important and has not been incorporated widely into the standard of care.

*****

Reviewer 2 Comments for the Author:

2.1. In this perspective piece, the authors discussed the potential role of acetaminophen in autism spectrum disorder (ASD) etiology based on a previous literature review.

This characterization of our perspective is inaccurate. The perspective piece is based on three recent reviews [2–4] as well as a systematic review of all original studies of acetaminophen safety in babies and children [1], not a single review. It should be pointed out that the three reviews each contained a detailed analysis of some aspects of previous studies connecting the pathology of ASD with acetaminophen.

2.2. The authors suggested that there is compelling evidence linking the use of acetaminophen to an increased ASD risk. The authors stated that acetaminophen is, "with no reasonable doubt," causally associated with ASD risk, acetaminophen is a necessary cause of ASD and "acetaminophen accounts for the vast majority of ASD cases."

This comment verifiably distorts the wording used in the perspective, which was very carefully chosen. We have previously concluded "without reasonable doubt" that acetaminophen is a causal agent in the induction of "many if not most" cases of ASD. However, the view that acetaminophen is a causal agent in the vast majority of cases is carefully worded in the perspective as follows: "Indeed, the simplest explanation for all observations is that acetaminophen accounts for the vast majority of cases of ASD." The assertion, perhaps implied by this reviewer by accident, that we have concluded without reasonable doubt that "acetaminophen accounts for the vast majority of ASD cases," is false.

2.3. The authors also recommended "eliminating neonatal exposure to acetaminophen" as an actionable intervention.

This statement is correct. We very strongly stand by that recommendation. The known toxicology and pharmacology of acetaminophen during the first days of human life were not considered when the drug was introduced into the pediatric population, and should be sufficient by themselves to make this recommendation. The fact that the anticipated adverse neurodevelopmental effects of the drug during this time period are evident (cord blood metabolite and circumcision associations with ASD) should provide overwhelming impetus to make this change in medical practice. The weight of 22 lines of evidence pointing at neuro-developmental toxicity of acetaminophen might also be considered, but seems unnecessary.

2.4. While I am deeply concerned with the previous research suggesting an association between acetaminophen and ASD risk, the authors make strong statements that may not be supported by the provided evidence.
It is crucial to exercise caution when interpreting the findings as they are not sufficient to conclude a direct causal relationship.
It is reasonable to disagree with the conclusions our team has reached, but only if the information is presented which provides a basis for that disagreement. As noted below, no valid basis for the reviewer's disagreement is provided.

2.5. The authors failed to address the limitations of the published studies.
This statement is verifiably false. Our recent reviews on this topic provide in-depth analyses of several lines of evidence linking ASD with acetaminophen use. These in-depth analyses provide extensive assessment of the limitations and weaknesses of those studies. Further detail is provided below under comments in which the reviewer elaborates on this opinion.

2.6. Some were based on animals, some were correlational, and some were observational and could be subject to various sources of bias, such as confounding and misclassification biases. For instance, since acetaminophen is an OTC drug, most of the available data sources do not accurately capture its usage. None of these designs are sufficient for establishing causation.
With this paragraph, the reviewer dismisses the evidence upon which we have based our previously published conclusions. Two main points should be considered in this regard: First, we do fully understand that

each study has weaknesses, and that no single study is alone sufficient to demonstrate causality. It is evident in our previous work that it is the weight of total evidence, not any one study, that carries sufficient strength to draw conclusions without reasonable doubt. The specific limitation mentioned by the reviewer is telling: it is true that OTC use (and other factors associated with faulty reporting of acetaminophen use) have limited the utility of some of the epidemiological studies. We conducted an in-depth analysis of this fact in two of our reviews [2, 4]. At the same time, compelling studies in animal models as well as several other lines of evidence (e.g., the study by Schultz in 2008 critically assessed in one of our recent reviews [2], the otherwise unexplained association between ASD and circumcision cited in all of our recent reviews [2–4] either capture OTC use or avoid OTC use as a complication).

The second point to consider in regard to this comment is that the evidence mentioned by the reviewer is, when considered objectively, overwhelming. For example, the reviewer implies that "animal studies" are not to be relied upon. However, as we have previously pointed out in one of our reviews, animals tend to be less sensitive to neurodevelopmental toxins, not more sensitive, than humans [4]. Further, as the neuroscientist on our team has previously pointed out in another of the reviews, animal models of neurodevelopment are widely accepted as reliable because neural development is so well conserved in mammals [3]. For these and other reasons, including the phylogenetically conserved metabolic pathways of acetaminophen, current evidence from animal studies should be highly regarded. Further, current FDA guidelines dictate that the results of these studies would absolutely preclude the use of acetaminophen during early brain development in humans if acetaminophen were put through the modern drug pipeline.

2.7. The piece relies on evidence from previous reviews by the authors on the association between acetaminophen and ASD. The search and selection criteria are not clear but it appears the authors only included studies with positive findings and did not review literature that did not report significant associations. Those pieces of work are expected to be underrepresented as they are affected by publication bias. Therefore, it is important to take this into consideration.

This statement is grossly misleading. Our most recent review [3] compiled 22 lines of evidence, only four of which involve direct associations between acetaminophen and ASD. Here the reviewer asserts that it is "expected" that other work showed no association, but was never published. However, questions regarding the impact of acetaminophen on neurodevelopment have been under scrutiny for more than a decade, and it seems unlikely that any proof of safety would go unpublished. It is true that, for reasons we have previously analyzed in depth in one of our reviews [4], many studies show ambiguous results due to a lack of statistical power. In that case, it is appropriate not to publish the study. (If the study conclusion is that acetaminophen might reduce the risk of ASD by 3% or it might increase the risk of ASD by 2-fold, it does not contribute to the literature because it is not informative for or against the risks of using acetaminophen.) That being said, the largest database (the DNBC) containing data on acetaminophen use and ASD has been analyzed, and the results have been published (providing one line of evidence out of 22 in our reviews). The results from this analysis are described below under point 2.9 of this review.

2.8. It is essential to note that even if such associations are valid, they do not necessarily imply a causal link, nor does it mean that acetaminophen is the most significant contributor to ASD risk.
Although technically correct, this statement is grossly misleading in this context. With a century of combined scientific experience among the authors of this perspective, we do understand that association does not equal causation. Again, our conclusions are based on 22 lines of evidence, many of which are not associations. This fact was evident in the perspective.

2.9. In fact, the small magnitude of association found in the Danish study (OR 1.3) suggests that acetaminophen alone cannot explain the majority of ASD cases, which contradicts the author's statement that "acetaminophen accounts for the vast majority of ASD cases." Such a modest risk increase in an observational study could be explained by confounding.
The reviewer's comment here reveals (a) a lack of appreciation for the limitations of the "Danish study," which did find an OR of 1.3, and

(b) a lack of knowledge regarding the evidence leading us to conclude that ASD is induced by acetaminophen. First, a detailed analysis of the results from the DNBC were presented in a previous study, and include a detailed assessment of the potential role of "confounding factors" [4]. In short, the Danish data indicate that only 7.7% of the mothers gave their child acetaminophen, which reflects dramatic under-reported based on studies which more specifically address acetaminophen use. This fact was emphasized in one of our recent reviews [4]. This reporting of acetaminophen use in the Danish data is probably less than about 10% of the actual use, based on our systematic review of the topic [2] and another study focused on acetaminophen in the Danish population [4]. An odds ratio of 0.3 for only 10% or less of the actual usage is wildly unacceptable and entirely consistent with our main conclusion. It is important to keep in mind that an assessment of the same database showed more than a 2-fold increase in infantile ASD with male circumcision, a procedure associated with acetaminophen use. Of equal importance is the fact that "confounding factors" considered in such studies, often direct or indirect reasons for giving acetaminophen, are known accelerants of acetaminophen toxicity because of the role of oxidative stress in the toxicity of that drug. This fact was emphasized in all three of our recent reviews on the topic [2–4]. Finally, and much more importantly, conclusive evidence is based on multiple lines of evidence that converge on a single conclusion, not upon a single line of evidence.

2.10. The figure demonstrates that all the environmental/genetic factors collectively are not sufficient causes of ASD; acetaminophen use is a necessary cause of ASD. This is not supported by the evidence provided by the authors.

The figure illustrates (not demonstrates) the model that we have concluded is correct. Stating that the model is not supported by the evidence should, again, have an underlying scientific basis. As discussed above, no valid basis for rejecting the model is provided by the reviewer.

2.11. Recommending "eliminating neonatal exposure to acetaminophen" is not warranted and is not feasible.

This statement is false. The relevant facts are that (a) available evidence demonstrates that the period of greatest risk for acetaminophen-mediated injury is within the first few postnatal hours and days [3], and (b) no study has ever shown that acetaminophen use during this period has long-term benefits. For this reason, the proposed change is quite reasonable. Please see also arguments under Point 2.3 from this reviewer.

2.12. As noted above, the evidence isn't sufficient to conclude a causal relationship or that the majority of ASD cases are attributed to acetaminophen.

This is the third time the reviewer has asserted that our conclusion is wrong. Repeated assertions of an unsupported statement do not constitute a valid review.

2.13. Until we confirm the causal mechanisms and magnitude of these risks in humans, weighing benefits and risks should always be the recommended approach.

The question here is obviously not whether it is recommended to weigh the benefits and risks before administering a drug, but rather what risks use of this particular drug actually poses. We have provided a litany of evidence in our previous reviews [2–4] upon which our perspective is based. Those reviews consider weaknesses and strengths of various lines of evidence, as well as the combined weight of the evidence. This consideration leads to our conclusion, without reasonable doubt, that acetaminophen exposure causes "many if not most cases" of ASD. As pointed out in the perspective, the simplest explanation for all observations is more dire: it is plausible that acetaminophen is responsible for the vast majority of all cases of ASD.

1. Cendejas-Hernandez J, Sarafian J, Lawton V, Palkar A, Anderson L, Lariviere V, et al. Paracetamol (Acetaminophen) Use in Infants and Children was Never Shown to be Safe for Neurodevelopment: A Systematic Review with Citation Tracking. *Eur J Pediatr.* 2022;181:1835–57. doi: 10.1007/s00431-022-04407-w.

2. Zhao L, Jones J, Anderson L, Konsoula Z, Nevison C, Reissner K, et al. Acetaminophen causes neurodevelopmental injury in susceptible babies and children: no valid rationale for controversy. *Clinical and Experimental Pediatrics.* 2023. Epub 2023/06/16. doi: 10.3345/cep.2022.01319. PubMed PMID: 37321575.

3. Parker W, Anderson LG, Jones JP, Anderson R, Williamson L, Bono-Lunn D, et al. The Dangers of Acetaminophen for Neurodevelopment Outweigh Scant Evidence for Long-Term Benefits. *Children.* 2024;11(1):44. PubMed PMID: doi:10.3390/children11010044.

4. Patel E, Jones JP, 3rd, Bono-Lunn D, Kuchibhatla M, Palkar A, Cendejas Hernandez J, et al. The safety of pediatric use of paracetamol (acetaminophen): a narrative review of direct and indirect evidence. *Minerva Pediatrics.* 2022;74(6):774–88. Epub 2022/07/14. doi: 10.23736/s2724-5276.22.06932-4. PubMed PMID: 35822581.

SECOND ROUND OF REVIEWS AND REBUTTALS FROM *PEDIATRICS*

The unedited correspondence from William Parker to the editor, including the letter to the editor, the reviewer comments (bold font), and the responses to the reviewer comments, is included in its entirety.

Dear Professor Kemper,

On behalf of all authors, I offer our sincere thanks for reconsidering the perspective and for securing an additional review. While this new reviewer is adamant that our conclusions are wrong, he/she does not provide any evidence suggesting that we could possibly be wrong, and provides substantial disinformation in the review. In terms of specific objections to the evidence provided, the reviewer's points fall into one of three categories: (A) verifiably false, (B) dependent on unorthodox arguments which are not consistent with known scientific principles, and (C) statements of fact that are not pertinent to the discussion. I provide a point-by-point rebuttal to the reviewer's comments below, including citations. Here I will provide a synopsis with some examples.

In total, the reviewer isolates 4 out of 22 lines of evidence, disputing those four lines of evidence. The remaining 18 lines of evidence are not considered. Three of those 4 lines of disputed evidence relate to the metabolism of acetaminophen, as outlined below in specific comments to the review. The reviewer contends that, based on metabolic considerations, children should be more resistant to acetaminophen-induced injury than are adults. As we have previously demonstrated, the widespread belief that acetaminophen is safe for neurodevelopment was based on flawed studies in the 1960s and 1970s which assumed that babies and small children metabolize acetaminophen the same as adults [1]. Further, the reviewer is verifiably misinformed regarding the metabolism of acetaminophen during development, as pointed out below in specific responses to the review. More concerning is that the reviewer ignores direct, repeated, and independent measurements of problems with drug metabolism (related to acetaminophen metabolism) in children with ASD.

The only remaining line of evidence disputed by the reviewer is a compelling line of evidence, but was inexplicably questioned because the initial intent of the study was not to evaluate a connection between acetaminophen and autism. Serendipity often plays a role in science and in medicine, and to void the results of any study because of the original intent of investigators unnecessarily and unconventionally restricts the potential impact of the study. Further, a similarly compelling study performed with intent to check associations between ASD and acetaminophen use was inexplicably ignored by the reviewer.

The reviewer objects a total of 8 times to our citations of the extensive reviews we have previously published on this topic. However, we are limited by the journal to only 7 citations regarding this very complex topic, necessitating our approach of citing previous reviews. If others had effectively reviewed the topic recently, we would cite those in lieu of our own reviews. Unfortunately, others have not reviewed this topic recently. Given the importance of this topic, it seems unwise to wait for others to review the topic before allowing the flagship journal of the American Academy to consider the conclusions we have reached.

We request that, based on the fact-based rebuttal below, you will reconsider your decision to reject the perspective. We would also request that any further reviews on this vitally important topic again be subject to fact-checking.

We thank you again for giving us the opportunity to submit our work to *Pediatrics*.

Sincerely,

William

William Parker, PhD

Visiting Scholar, Department of Psychology and Neuroscience,

University of North Carolina, Chapel Hill CEO, WPLab, Inc., a 501(c)3 nonprofit conducting research and education on the cause of chronic inflammatory disease in high income areas of the world

Associate Professor, Duke University, retired

Critique of third Reviewer Comments:
Page 4, first section: (the widely accepted "multifactorial model"): there are no references for this paragraph.
The "multifactorial model" is common knowledge in the field. We are limited to only 7 references due to the formatting requirements of the journal. Given that the number of references is severely restricted, we have avoided citation of information that is common knowledge.

P4, line 53: There are no references or data available for this literature search.
We are able to provide a supplemental data file containing the details of the literature search, including all references, if the journal will allow us to attach the document to supplemental information. However, details regarding the search are provided so that it can be easily reproduced, and the results are not surprising to investigators in the field.

P4, line 60–62: The sentence is misleading. It says the evidence has been reviewed, glossing over the fact that the cited evidence (refs 1–4) was generated by the first author on the paper. A phrase like "Building on our previous work," would be more clear, though to meet the scientific rigor of the journal, the authors should be able to cite a much larger body of references, from various authors and credible journals.
The reviewer raises two issues here, one of which is misleading, and the other of which is false. First, we agree that it is best to revise the manuscript, indicating that we are the ones who have previously reviewed this topic. However, our group did not "generate" the evidence. The vast majority of the evidence we reviewed is from other laboratories. Second, we strongly disagree with the assertion that scientific rigor is in any way demonstrated by consensus. Consensus without evidence is only agreement, not the systematic endeavor of science.

P4 line 63: "no reasonable doubt" is not widely accepted by the medical community. The "20 lines of evidence" refer to 3 papers written by the first author.
We understand that the conclusions we have reached are not currently accepted by the medical community. However, the absence of acceptance

does not negate the validity of our conclusions. It goes without saying that any innovation in medicine does not necessarily start immediately with wide-spread acceptance. We welcome the opportunity to consider evidence inconsistent with our conclusions. This reviewer provides none.

P4 line 64: "Among the evidence lies over a dozen associations"—it is unclear what is meant by this.
This statement is difficult to understand since the next sentence of the paragraph provides two specific examples, and the reviews cited provide detailed lists with references.

P4, line 66: "examples of observations that lack any reasonable explanation other than the involvement of acetaminophen in the pathogenesis of ASD include the association of ASD with... [ ] ... circumcision"—this sentence seems to be from reference 3, which is cited for the previous sentence. If you go to reference 3, it is a review paper by the same authors that summarizes another paper: "Circumcision of males, often performed using APAP as an analgesic, is associated with a two-fold increase in the risk for early-onset (infantile) ASD.' However, this paper was not designed to investigate the impact of APAP on circumcision and ASD. The authors of that paper conclude: "Unfortunately, we had no data available on analgesics or possible local anaesthetics used during ritual circumcisions in our cohort, so we were unable to address the paracetamol hypothesis directly."
The striking connection between ASD and circumcision is listed as one of two examples (among the list of over a dozen examples previously published) of *associations* that lack explanations other than the causal role of acetaminophen in ASD, as indicated in the paragraph. It is correct that the authors of the study investigating the association between circumcision and ASD do not have data regarding acetaminophen use in the procedure, but the drug is commonly used in the procedure, and the association is explained by the model presented. Further, although the authors could not obtain records of acetaminophen use during circumcision, the reason they performed the study was that another investigator, Ann Bauer, as previously hypothesized that acetaminophen use during circumcision was a cause of some cases of ASD [2]. Thus, the fact that

the study in question describes an association does not negate the observation or consideration of that observation when drawing conclusions.

Notably, the reviewer did not comment on the cord-blood association between acetaminophen and ASD [3], which did directly measure acetaminophen and which was mentioned in the manuscript. Of the two examples provided in the manuscript, the reviewer objected to the indirect association, but did not mention the direct association. Nevertheless, both associations, and well as numerous others described in the references cited, all support the same conclusion.

P5, line 73: "This discrepancy, referred to as "missing heritability," is resolved by the fact that siblings and particularly twins are expected to have . . ." It appears as though this is association and not causation. We understand that association does not equal causation. At the same time, when multiple associations are all explained by a single cause, and when other lines of evidence, including work in animal models and plausible mechanisms, support the conclusion of causation, then that conclusion should be reached. We are careful in our wording to describe when an association exists versus when causation has been demonstrated, usually in animal studies with rigorous controls. However, at the same time, it is legitimate to state than an association is consistent with causation.

P5, line 80: "A tally published in 2022"—consider replacing with "We previously report"; also, the sentence is very unclear—what actual research evidence is there that supports this? Perhaps a chart could help? Consider citing human studies and other credible sources in which your group are not the authors. For this citation, the cited reference describes APAP metabolism and NAPQI generation (correctly), but then asserts that children in utero experience GSH depletion. Again, the citation for this sentence is the authors own previous work. They go on to add a host of conditions with no citation and no evidence to support that these conditions themselves lead to GSH depletion. Moreover, these populations are not widely viewed as being at increased risk of APAP toxicity and are not widely accepted conditions associated with GSH depletion.

As mentioned above, we agree that it is best to revise the manuscript, indicating that we are the ones who have previously reviewed this topic.

The entirety of the sentence that the reviewer finds unclear is "A tally published in 2022 (citation here) found six unknown factors that must be invoked and eight largely independent observations that must be attributed to coincidence if acetaminophen is not a neurodevelopmental toxin." The sentence refers to a previous review, published by us, which lists the evidence in detail, the large majority of which was not derived from our laboratory.

We do understand that the populations mentioned are not considered to be at risk for acetaminophen-induced injury. Nevertheless, as explained in the study cited, oxidative stress will drive the equilibrium of acetaminophen metabolism toward the toxic metabolite. GSH was not mentioned in the article. (GSH depletion is only one of three mechanisms by which the amount of toxic metabolite can be increased.) Fortunately, studies in children with ASD have been conducted, and show impaired ability to metabolize acetaminophen, which does in fact increase the amount of oxidized product (toxic metabolite) formed.

P5, line 90: The authors cite their own work. I am interested in a sulfation deficiency hypothesis, but in fact, children in general have a higher capacity to sulfate and therefore are generally relatively protected from APAP toxicity, including APAP-induced hepatotoxicity. Also, cited reference #7 is a pilot study, and was designed to study parent claims that certain foods are associated with an increase in autistic behaviors, using paracetamol as a marker of sulfation and gluronidation. It was not designed to determine whether children with ASD metabolize the drug efficiently.

As mentioned above, we agree that it is best to revise the manuscript, indicating that we are the ones who have previously reviewed this topic. That being said, the actual information cited in our previous reviews is, for the most part, from other laboratories. Thus we cannot claim the we are citing our work, but rather that we are citing our review of work by others.

For example, here the reviewer asserts that children in general have a "higher capacity" to sulfate, and are therefore resistant to acetaminophen-induced injury. As we have previously noted [4, 5], the most sensitive time of life to acetaminophen-induced injury is during the neonatal period. Studies of drug metabolism in neonates have never indicated that these young individuals have *better* sulfation reactions. These studies indicate that the *only* significant means by which neonates can metabolize acetaminophen via phase II pathways is through sulfation, not that they have enhanced sulfation. (Sulfation is near adult levels in neonates, whereas glucuronidation, the other major pathway for phase II metabolism of acetaminophen, is not.) For example, the biologic half-life of acetaminophen is 3.5 hours in neonates as compared to about 2 hours in adults [6], indicating that the total capacity to metabolize acetaminophen per unit weight is less in neonates than in adults. Equally important is the observation that the relative fraction of acetaminophen that is oxidized into the toxic metabolite is almost 20% in 7.5 day old neonates [7], more than double the percent oxidized in adults [8]. More importantly is the considerable body of literature indicating that children with ASD have impaired sulfation reactions.

More importantly, sulfation in children with ASD is impaired based on multiple, independent studies, as recently reviewed by Richard Williams [9]. These studies all show that children with ASD tend to have problems with sulfation, which is known to be the primary path used by neonates to detoxify acetaminophen. We understand that the intent of reference #7 was not to identify whether or not children with ASD can metabolize acetaminophen efficiently. However, the authors clearly did find that children with ASD cannot metabolize acetaminophen efficiently. It is not rationale to ignore the results of a study because the original intent of the study was not related to the implications of the study.

Although neonates are relatively protected from acetaminophen-induced hepatotoxicity, available evidence indicates that the CNS rather than the liver is the target of injury in susceptible neonatal individuals. For example, studies in laboratory animal models as early as 1984 demonstrated that lethal doses of acetaminophen do not cause hepatotoxicity in neonates [10]. Further, as pointed out by William L. Nyhan, the noted pediatrician/scientist, differences between neonates and older children tend to render neonates more sensitive to numerous drugs, not less sensitive [11].

P6, line 93: The cited reference #3 is authors own work and appears to be a review paper (no methods listed). Please cite primary references to support this claim.

We are limited to only 7 references due to the formatting requirements of the journal. We would be able to add more references if the format of the article were changed to a review, for example. However, the assertions made on P6, line 93 are fully supported in the cited reference.

P6, line 96: The authors state here that the "multifactorial model is critical part of the picture" which seems to contradict the Occam's razor example on the previous page.

Occam's razor dictates that the simplest answer *which accounts for available data* is likely the correct one. It does not say that the simplest answer is likely the correct one. This fact is stated clearly in the manuscript. Thus, our conclusions are strongly supported by Occam's razor, as described in the manuscript.

P6, line 103: Regarding factors associated with ASD that can cause oxidative stress- I have not seen evidence that sex, Down syndrome, preterm birth, high vitamin B levels, leads to elevated NAPQI production and do not recommend alternate dosing or management of APAP exposures in these patients. The cited reference appears to show only a few citations that show association, not causation. The cited reference for impaired APAP clearance in unconjugated hyperbili pts is for repeat dose IV APAP, not maternal use or PO APAP, and is an association.

The reference cited here [12] is an extensive review of acetaminophen metabolism, and describes almost 30 associations between ASD and factors associated with oxidative stress and/or inflammation. Describing the manuscript, with over 30 pages with almost 200 citations, as having "only a few citations that show association" may suggest some bias on the part of the reviewer. We do understand that the formation of NAPQI has not been directly measured with most of those associations, but acetaminophen metabolism is among the most widely studied drug metabolism in history. Oxidative stress is expected to drive enhanced production of the toxic intermediate based on current understanding of the metabolism of that drug. We understand that the information is an association only. However, the almost 30 associations only account for one of 22 lines of evidence, and demonstrate that the mechanism of drug injury is plausible.

The reviewer objects to the conditions in which one study found that acetaminophen was poorly metabolized by children with ASD compared to controls. However, the point here is the difficultly with metabolism, not the conditions under which the study was performed. Further, as pointed out above, that finding has been supported by multiple, independent studies, as reviewed by Richard Williams [9].

P6, line 112: This seems dangerously overstated and not supported by research evidence.
We understand that the reviewer disagrees with us, but at the same time, the reviewer does not provide any evidence that our conclusions are incorrect. We argue that to ignore the overwhelming evidence in this case is exceedingly dangerous.

P7, line 122: The sentence starting with "For example, many if not most . . ." is difficult to understand. Recommend rewording to make it clearer and more easily readable, using primary references that are not wholly comprised of the authors' own work.
The full statement which the reviewer finds difficult to understand is "For example, many if not most current medical practices resulting in exposure of the developing brain to acetaminophen are either (a) not currently recommended, (b) shown to be ineffective by published medical studies, or (c) have never been shown to have long-term benefits by published medical studies (citation)." To improve clarity, we could revise the sentence to read "For example, many if not most current medical practices that result in exposing the developing brain to acetaminophen are (a) not currently recommended, (b) shown to be ineffective by published medical studies, or (c) have never been shown to have long-term benefits by published medical studies (citation)."

Again, we are limited to only 7 references due to the formatting requirements of the journal. However, the assertions made on P7, line 122 are fully supported in the cited reference.

P8, line 139: recommend changing out "non-sequitur" as some view that it does logically follow the available evidence and is not consistent with how the word is used colloquially.

The term "non-sequitur" is used correctly in this sentence. The validity of the multifactorial model of ASD pathogenesis should not be taken to mean that no single factor can induce the vast majority of ASD.

P8, line 139: I disagree that there is overwhelming evidence. Most of the cited references are reviews. Most of the cited references are authors' own work.

The fact that we cite extensive reviews of our own work is an artifact of the manuscript format, not any indication of an absence of evidence. The cited references tend to be reviews because we are limited to 7 references by the journal formatting. Of the 22 lines of evidence, only two lines include any of our group's work, and only one of those lines of evidence is derived exclusively from our group: the formal proof that acetaminophen was never shown to be safe for pediatric use.

P8, line 143: I wish this were true, but this conclusion is overreaching ("clear path") and not clearly supported by the available evidence and citations in this paper.

We understand that the reviewer disagrees with us, but at the same time, the reviewer does not provide any evidence that our conclusions are incorrect. Perhaps more importantly, the reviewer's objections are based largely on verifiably false assertions, which include the ideas that scientific rigor is based on consensus, our conclusions do not satisfy Occam's razor, and neonates are more resistant to drug-induced injury than adults. Further, the reviewer takes the position that findings in studies not deliberately designed to determine the impact of acetaminophen exposure on neurodevelopment, but yet yielding compelling evidence relevant to the question, are not applicable. This attitude is not justifiable. Discovery in the fields of medicine and science are replete with serendipity, and it is to our own detriment to disregard it.

1. Cendejas-Hernandez J, Sarafian J, Lawton V, Palkar A, Anderson L, Lariviere V, et al. Paracetamol (Acetaminophen) Use in Infants and Children was Never Shown to be Safe for Neurodevelopment: A Systematic Review with Citation Tracking. *Eur J Pediatr.* 2022;181:1835–57. doi: 10.1007/s00431-022-04407-w.

2. Bauer A, Kriebel D. Prenatal and perinatal analgesic exposure and autism: an ecological link. *Environmental Health.* 2013;12(1):41. PubMed PMID: doi:10.1186/1476-069X-12-41.

3. Ji Y, Azuine RE, Zhang Y, Hou W, Hong X, Wang G, et al. Association of Cord Plasma Biomarkers of In Utero Acetaminophen Exposure With Risk of Attention-Deficit/Hyperactivity Disorder and Autism Spectrum Disorder in Childhood. *JAMA Psychiatry.* 2020;77(2):180–89. Epub 2019/10/31. doi: 10.1001/jamapsychiatry.2019.3259. PubMed PMID: 31664451; PubMed Central PMCID: PMCPMC6822099.

4. Patel E, Jones JP, 3rd, Bono-Lunn D, Kuchibhatla M, Palkar A, Cendejas Hernandez J, et al. The safety of pediatric use of paracetamol (acetaminophen): a narrative review of direct and indirect evidence. *Minerva Pediatrics.* 2022;74(6):774–88. Epub 2022/07/14. doi: 10.23736/s2724-5276.22.06932-4. PubMed PMID: 35822581.

5. Parker W, Anderson LG, Jones JP, Anderson R, Williamson L, Bono-Lunn D, et al. The Dangers of Acetaminophen for Neurodevelopment Outweigh Scant Evidence for Long-Term Benefits. *Children.* 2024;11(1):44. PubMed PMID: doi:10.3390/children11010044.

6. Levy G, Khanna NN, Soda DM, Tsuzuki O, Stern L. Pharmacokinetics of acetaminophen in the human neonate: formation of acetaminophen glucuronide and sulfate in relation to plasma bilirubin concentration and D-glucaric acid excretion. *Pediatrics.* 1975;55(6):818–25. Epub 1975/06/01. PubMed PMID: 1134883.

7. Cook SF, Stockmann C, Samiee-Zafarghandy S, King AD, Deutsch N, Williams EF, et al. Neonatal Maturation of Paracetamol (Acetaminophen) Glucuronidation, Sulfation, and Oxidation Based on a Parent-Metabolite Population Pharmacokinetic Model. *Clin Pharmacokinet.* 2016;55(11):1395–411. Epub 2016/10/21. doi: 10.1007/s40262-016-0408-1. PubMed PMID: 27209292; PubMed Central PMCID: PMCPMC5572771.

8. Mazaleuskaya LL, Sangkuhl K, Thorn CF, FitzGerald GA, Altman RB, Klein TE. PharmGKB summary: pathways of acetaminophen metabolism at the therapeutic versus toxic doses. *Pharmacogenet Genomics.* 2015;25(8):416–26. Epub 2015/06/08. doi: 10.1097/fpc.0000000000000150. PubMed PMID: 26049587; PubMed Central PMCID: PMCPMC4498995.

9. Williams RJ. Sulfate Deficiency as a Risk Factor for Autism. *J Autism Dev Disord.* 2020;50(1):153–61. Epub 2019/09/29. doi: 10.1007/s10803-019-04240-5. PubMed PMID: 31562579; PubMed Central PMCID: PMCPMC6946761.

10. Green MD, Shires TK, Fischer LJ. Hepatotoxicity of acetaminophen in neonatal and young rats. I. Age-related changes in susceptibility. *Toxicol Appl Pharmacol.* 1984;74(1):116–24. Epub 1984/06/15. doi: 10.1016/0041-008x(84)90277-1. PubMed PMID: 6729816.

11. Nyhan WL. Toxicity of drugs in the neonatal period. *J Pediatr.* 1961;59:1–20. Epub 1961/07/01. doi: 10.1016/s0022-3476(61)80204-7. PubMed PMID: 13729971.

12. Parker W, Hornik CD, Bilbo S, Holzknecht ZE, Gentry L, Rao R, et al. The role of oxidative stress, inflammation and acetaminophen exposure from birth to early childhood in the induction of autism. *J Int Med Res.* 2017;45(2):407–38.

ACKNOWLEDGMENTS

Scientists depend on their peers. In one example, my colleagues eliminated laboratory mouse pups the size of medium-sized elephants and carbon atoms with twelve protons, two equally absurd apparitions that appeared in the first draft of this book. Without the labor of love from many colleagues in pursuit of a better understanding of autism spectrum disorders, this book would not have been possible.

I owe a great debt of gratitude to Susanne Meza-Keuthen, my ever-patient wife, whose love and support are so incredible. I greatly value the prayers of my mother, Frances Carolyn Parker, for me and for this book. She and my father instilled in me a strong work ethic that Pill Song, my PhD advisor, effectively reinforced. Without the strong start that Pill gave me in graduate school, my academic career would never have been possible.

I am very grateful to those who fostered my love of science and math, from middle school to university. Morris Conrad, Sandra Sawyer-Mobley, and Robert Steinmeier loved learning in a way that was contagious.

The creativity and imagination of my postdoctoral advisor, Jeffrey Platt, were inspirational. Jeff values ideas, and when I had a great idea, he wanted no credit for it. For ideas I developed while working for him, he ensured that I received all the credit, to the point of refusing even an acknowledgment on the manuscripts describing the idea. This sort of commitment to a trainee's career is truly exceptional, and is unique in the world of science in my experience. I owe him a great debt of thanks for that. Fortunately, he can't stop me from acknowledging him in this book.

While at Duke, I had the great pleasure of working with Mary Lou Everett and Zoie Holzknecht, professional scientists who were technically gifted. They conducted meticulous experiments in the laboratory with me literally for decades, helping to train students and taking care of laboratory administration. Without their efforts, we would never have uncovered the mysteries of the human gut that eventually led to our work on autism. I am also grateful to Jasmine Cendejas-Hernandez, who conducted some of the most tedious work imaginable in our efforts to understand the connection between acetaminophen and autism.

I owe a great debt of thanks to many individuals who held the research together as I left Duke. Tabitha Parker, John Poulton, and Susan Poulton provided incredible and absolutely critical support. Randy Bollinger, one of the kindest people I have ever known, provided wonderful scientific support and encouragement as I left Duke. J. P. Jones and Roula Konsoula continued to volunteer on our projects, applying their invaluable knowledge in the field of drug design and development to our work on acetaminophen.

Kate Reissner is objective, brilliant, and most importantly, a person of unimpeachable integrity. When she accepted me into her laboratory at the University of North Carolina in Chapel Hill, she effectively restored my connection to academia, erasing any limitations on my scientific work that my departure from Duke had imposed.

Lauren Williamson was and still is a wonderful collaborator, before and after I left Duke. In a world where many faculty know more about the deadlines for funding application cycles than about the scientific instrumentation in their own laboratory, Lauren Williamson knows the nuts and bolts of everything in her laboratory, and she thinks about what's important rather than how to follow the latest trends in her field for the sake of funding. She is a master of running wonderful experiments on a shoestring budget.

I am grateful to Paul Corrigan, a wordsmith extraordinaire, for crafting one of our papers for us. He took the time to understand our work and to guide us in presenting that work in an elegant manner.

Others, including Christina Flynn, Devora Buechler, Rachel Anderson, and Gina Vitale, provided both emotional and intellectual support as I left Duke. Nathan Kohuth and Spencer Sharpe took care of our logistics and administration, effectively handling much of the work that had been performed by Duke administration.

After leaving Duke, I participated in wonderful interviews with Len Arcuri, Brian Hooker, Vanessa Lopez, Laura Loomer, Sheldon Baker, Marta Zaraska, Tim Gabor, Nick Jikomes, Chris Magryta, Elanie Welch, and Honey Rinicella, and others. These individuals were interested in the science connecting acetaminophen and autism, and took that time to listen to the evidence and to publicize that evidence.

I owe a debt of thanks to dozens of others, including the research team at the University of North Carolina in Chapel Hill, who have participated in many productive discussions and helped in many ways support our work probing the connection between acetaminophen and autism.

Finally, without the work from dozens of laboratories around the world whose publications are cited in this book, our understanding of the connection between acetaminophen and autism would not be possible.

NOTES

Introduction

1   Ahlqvist, V.H., Sjöqvist, H., Dalman, C., et al., "Acetaminophen Use During Pregnancy and Children's Risk of Autism, ADHD, and Intellectual Disability," *JAMA* 331, no. 14 (2024): 1205–14. https://doi.org/10.1001/jama.2024.3172.

2   Prada, D., Ritz, B., Bauer, A.Z., and Baccarelli, A.A., "Evaluation of the Evidence on Acetaminophen Use and Neurodevelopmental Disorders Using the Navigation Guide Methodology," *Environmental Health* 24, no. 1 (2025): 56. https://doi.org/10.1186/s12940-025-01208-0.

3   Yousif, S.R., Aboody, R., and Keil, F.C., "The Illusion of Consensus: A Failure to Distinguish between True and False Consensus," *Psychological Science* 30, no. 8 (2019): 1195–204. https://doi.org/10.1177/0956797619856844.

4   "Baylor College of Medicine. Dr. Klotman's Video Message – Week 293." https://www.youtube.com/watch?v=Xnc2i2_R2MA.

5   Parker, W., Corrigan, P.T., Anderson, R., et al., "Evidence That Acetaminophen Triggers Autism in Susceptible Individuals Has Been Ignored and Mishandled for More Than a Decade," *Journal of the Academy of Public Health* (2025). https://publichealth.realclearjournals.org/literature-syntheses/2025/10/evidence-that-acetaminophen-triggers-autism-in-susceptible-individuals-has-been-ignored-and-mishandled-for-more-than-a-decade/.

6   Swanson, E., "Peer Review, Confidential Comments to the Editor, and the Golden Rule," *Annals of Plastic Surgery* 90, no. 1 (2023): 1–3. https://doi.org/10.1097/sap.0000000000003320.

7   Parker, W., Corrigan, P.T., Anderson, R., et al., "Evidence That Acetaminophen Triggers Autism in Susceptible Individuals Has Been Ignored and Mishandled for More Than a Decade," *Journal of the Academy of Public Health* (2025). https://publichealth.realclearjournals.org/literature-syntheses/2025/10/evidence-that-acetaminophen-triggers-autism-in-susceptible-individuals-has-been-ignored-and-mishandled-for-more-than-a-decade/.

8   Parker, W., Corrigan, P.T., Anderson, R., et al., "Evidence That Acetaminophen Triggers Autism in Susceptible Individuals Has Been Ignored and Mishandled for More Than a Decade," *Journal of the Academy of Public Health* (2025). https://publichealth.realclearjournals.org/literature-syntheses/2025/10/evidence-that-acetaminophen-triggers-autism-in-susceptible-individuals-has-been-ignored-and-mishandled-for-more-than-a-decade/.

Chapter 1: The Tylenol-Autism Whisperer

9   Chen, E., Everett, M.L., Holzknecht, Z.E., et al., "Short-Lived Alpha-Helical Intermediates in the Folding of Beta-Sheet Proteins," *Biochemistry* (2010).
Parker, W., and Song, P.S., "Location of Helical Regions in Tetrapyrrole-Containing Proteins by a Helical Hydrophobic Moment Analysis. Application to Phytochrome," *Journal of Biological Chemistry* 265, no. 29 (1990): 17568–75.
Parker, W., and Song, P.S., "Protein Structures in SDS Micelle-Protein Complexes," *Biophysical Journal* 61, no. 5 (1992): 1435–39.
Parker, W., and Stezowski, J.J., "The Surface of Beta-Sheet Proteins Contains Amphiphilic Regions Which May Provide Clues About Protein Folding," *Proteins* 25, no. 2 (1996): 253–60.
Parker, W., Sood, A., and Song, A., "Organization of Regions with Amphiphilic Alpha-Helical Potential within the Three-Dimensional Structure of Beta-Sheet Proteins," *Protein Engineering* 14 (2001): 315–19.

10   Chen, E., Parker, W., Lewis, J.W., Song, P.-S., and Kliger, D.S., "Time-Resolved UV Circular Dichroism of Phytochrome A: Folding of the N-Terminal Region," *Journal of the American Chemical Society* 115, no. 21 (1993): 9854–55.

11   Stezowski, J.J., Parker, W., Hilgenkamp, S., and Gdaniec, M., "Pseudopolymorphism in Tetradeca-2,6-O-Methyl-B-Cyclodextrin: The Crystal Structures for Two New Hydrates—Conformational Variability in the Alkylated B-Cyclodextrin Molecule," *Journal of the American Chemical Society* 123, no. 17 (2001): 3919–26. https://doi.org/10.1021/ja002164l.

12   Parker, W., and Stezowski, J.J., "The Surface of Beta-Sheet Proteins Contains Amphiphilic Regions Which May Provide Clues About Protein Folding," *Proteins* 25, no. 2 (1996): 253–60.

13   Collins, B.H., Cotterell, A.H., McCurry, K.R., et al., "Cardiac Xenografts between Primate Species Provide Evidence for the Importance of the Alpha-Galactosyl Determinant in Hyperacute Rejection," *Journal of Immunology* 154, no. 10 (1995): 5500–10.
Cotterell, A.H., Collins, B.H., Parker, W., Harland, R.C., and Platt, J.L., "The Humoral Immune Response in Humans Following Cross-Perfusion of Porcine Organs," *Transplantation* 60, no. 8 (1995): 861–68.

14   Everett, M.L., Palestrant, D., Miller, S.E., Bollinger, R.B., and Parker, W., "Immune Exclusion and Immune Inclusion: A New Model of Host-Bacterial Interactions in the Gut," *Clinical and Applied Immunology Reviews* 5 (2004): 321–32.
Bollinger, R.R., Everett, M.L., Palestrant, D., Love, S.D., Lin, S.S., and Parker, W., "Human Secretory Immunoglobulin a May Contribute to Biofilm Formation

in the Gut," *Immunology* 109, no. 4 (2003): 580–87. https://doi
.org/10.1046/j.1365-2567.2003.01700.x.
Dishaw, L.J., Cannon, J.P., Litman, G.W., and Parker, W., "Immune-Directed
Support of Rich Microbial Communities in the Gut Has Ancient Roots,"
Developmental & Comparative Immunology 47, no. 1 (2014): 36–51. https://doi
.org/10.1016/j.dci.2014.06.011.

15 Bollinger, R.B., Barbas, A.S., Bush, E.L., Lin, S.S., and Parker, W., "Biofilms
in the Large Bowel Suggest an Apparent Function of the Human Vermiform
Appendix," *Journal of Theoretical Biology* 249 (2007): 826–31. https://doi
.org/10.1016/j.jtbi.2007.08.032.
Smith, H.F., Fisher, R.E., Everett, M.L., Thomas, A.D., Bollinger, R.B., and
Parker, W., "Comparative Anatomy and Phylogenetic Distribution of the
Mammalian Cecal Appendix," *Journal of Evolutionary Biology* 22 (2009):
1984–99. https://doi.org/10.1111/j.1420-9101.2009.01809.
Laurin, M., Everett, M.L., and Parker, W., "The Cecal Appendix: One More
Immune Component with a Function Disturbed by Post-Industrial Culture,"
Anatomical Record 294, no. 4 (2011): 567–79. https://doi.org/10.1002/ar.21357.
Smith, H.F., Parker, W., Kotze, S.H., and Laurin, M., "Multiple Independent
Appearances of the Cecal Appendix in Mammalian Evolution and an
Investigation of Related Ecological and Anatomical Factors," *Comptes Rendus
Palevol* (2013). https://doi.org/doi:10.1016/j.crpv.2012.12.001.

16 Collard, M.K., Bardin, J., Marquet, B., Laurin, M., and Ogier-Denis, É.,
"Correlation between the Presence of a Cecal Appendix and Reduced Diarrhea
Severity in Primates: New Insights into the Presumed Function of the
Appendix," *Scientific Reports* 13, no. 1 (2023): 15897. https://doi.org/10.1038/
s41598-023-43070-5.

17 Bollinger, R.B., Everett, M.L., Wahl, S., Lee, Y.-H., Orndorff, P.E., and Parker,
W., "Secretory Iga and Mucin-Mediated Biofilm Formation by Environmental
Strains of Escherichia Coli: Role of Type 1 Pili," *Molecular Immunology* 43
(2006): 378–87.

18 Lee, S.Y.R., and Parker, W., "Amphiphilic Alpha-Helical Potential: A Putative
Folding Motif Adding Few Constraints to Protein Evolution," *Journal of
Molecular Evolution* 73, no. 3–4 (2011): 166–80. https://doi.org/doi:10.1007/
s00239-011-9465-0.

19 Bollinger, R.B., Everett, M.L., Wahl, S., Lee, Y.-H., Orndorff, P.E., and Parker,
W., "Secretory Iga and Mucin-Mediated Biofilm Formation by Environmental
Strains of Escherichia Coli: Role of Type 1 Pili," *Molecular Immunology* 43
(2006): 378–87.

20 Palestrant, D., Holzknecht, Z.E., Collins, B.H., Miller, S.E., Parker, W., and
Bollinger, R.R., "Microbial Biofilms in the Gut: Visualization by Electron
Microscopy and by Acridine Orange Staining," *Ultrastructural Pathology* 28
(2004): 23–27.

21 Devalapalli, A.P., Lesher, A., Shieh, K., et al., "Increased Levels of IgE and
Autoreactive, Polyreactive IgG in Wild Rodents: Implications for the Hygiene
Hypothesis," *Scandinavian Journal of Immunology* 64 (2006): 125–36. https://
doi.org/10.1111/j.1365-3083.2006.01785.x.

Lesher, A., Li, B., Whitt, P., et al., "Increased Il-4 Production and Attenuated Proliferative and Proinflammatory Responses of Splenocytes from Wild-Caught Rats (Rattus Norvegicus)," *Immunology and Cell Biology* 84 (2006): 374–82.

Lin, S.S., Holzknecht, Z.E., Trama, A.M., et al., "Immune Characterization of Wild-Caught *Rattus Norvegicus* Suggests Diversity of Immune Activity in Biome-Normal Environments," *Journal of Evolutionary Medicine* 1 (2012): 1–16.

Trama, A.M., Holzknecht, Z.E., Thomas, A.D., et al., "Lymphocyte Phenotypes in Wild-Caught Rats Suggest Potential Mechanisms Underlying Increased Immune Sensitivity in Post-Industrial Environments," *Cellular & Molecular Immunology* 9 (2012): 163–74.

22   Cheng, A.M., Jaint, D., Thomas, S., Wilson, J., and Parker, W., "Overcoming Evolutionary Mismatch by Self-Treatment with Helminths: Current Practices and Experience," *Journal of Evolutionary Medicine* 3 (2015): Article ID 235910.

Liu, J., Morey, R.A., Wilson, J.K., and Parker, W., "Practices and Outcomes of Self-Treatment with Helminths Based on Physicians' Observations," *Journal of Helminthology* FirstView (2016): 1–11.

Williamson, L.L., McKenney, E.A., Holzknecht, Z.E., et al., "Got Worms? Perinatal Exposure to Helminths Prevents Persistent Immune Sensitization and Cognitive Dysfunction Induced by Early-Life Infection," *Brain, Behavior, and Immunity* 51 (2016): 14–28. https://doi.org/10.1016/j.bbi.2015.07.006.

Parker, W., "Not Infection with Parasitic Worms, but Rather Colonization with Therapeutic Helminths," *Immunology Letters* (2017). https://doi.org/http://dx.doi.org/10.1016/j.imlet.2017.07.008.

Sobotková, K., Parker, W., Levá, J., Růžková, J., Lukeš, J., and Jirků Pomajbíková, K., "Helminth Therapy—from the Parasite Perspective," *Trends in Parasitology* 35, no. 7 (2019): 501–15. https://doi.org/10.1016/j.pt.2019.04.009.

Venkatakrishnan, A., Sarafian, J.T., Jirků-Pomajbíková, K., and Parker, W., "Socio-Medical Studies of Individuals Self-Treating with Helminths Provide Insight into Clinical Trial Design for Assessing Helminth Therapy," *Parasitology International* 87 (2021): 102488. https://doi.org/10.1016/j.parint.2021.102488.

23   Liu, J., Morey, R.A., Wilson, J.K., and Parker, W., "Practices and Outcomes of Self-Treatment with Helminths Based on Physicians' Observations," *Journal of Helminthology* FirstView (2016): 1–11.

24   Parker, W., Sarafian, J.T., Broverman, S.A., and Laman, J.D., "Between a Hygiene Rock and a Hygienic Hard Place: Avoiding SARS-CoV-2 While Needing Environmental Exposures for Immunity," *Evolution, Medicine and Public Health* 9, no. 1 (2021): 120–30. https://doi.org/10.1093/emph/eoab006.

25   Wolday, D., Gebrecherkos, T., Arefaine, Z.G., et al., "Effect of Co-Infection with Intestinal Parasites on COVID-19 Severity: A Prospective Observational Cohort Study," *eClinicalMedicine* 39 (2021): 101054. https://doi.org/10.1016/j.eclinm.2021.101054.

Wolday, D., Tasew, G., Amogne, W., et al., "Interrogating the Impact of Intestinal Parasite-Microbiome on Pathogenesis of COVID-19 in Sub-Saharan Africa," *Frontiers in Microbiology* 12 (2021): 614522. https://doi.org/10.3389/fmicb.2021.614522.

26 Ren, L., Holzknecht, R.A., Holzknecht, Z.E., et al., "A Mole Rat's Gut Microbiota Suggests Selective Influence of Diet on Microbial Niche Space and Evolution," *Experimental Biology and Medicine (Maywood, N.J.)* 244, no. 6 (2019): 471–83. https://doi.org/10.1177/1535370219828703.

27 Jirků, M., Parker, W., Kadlecová, O., et al., "Developmental Plasticity Enables Intestinal Tapeworm to Adapt to Dietary Stress," *Nature Communications* (2026): https://doi.org/10.1038/s41467-026-69475-0

28 Engelenburg, H.J., Lucassen, P.J., Sarafian, J.T., Parker, W., and Laman, J.D., "Multiple Sclerosis and the Microbiota: Progress in Understanding the Contribution of the Gut Microbiome to Disease," *Evolution, Medicine, and Public Health* 10, no. 1 (2022): 277–94. https://doi.org/10.1093/emph/eoac009. Parker, W., Patel, E., Jirků-Pomajbíková, K., and Laman, J.D., "COVID-19 Morbidity in Lower Versus Higher Income Populations Underscores the Need to Restore Lost Biodiversity of Eukaryotic Symbionts," *iScience* 26, no. 3 (2023): 106167. https://doi.org/10.1016/j.isci.2023.106167. Parker, W., Jirků, K., Patel, E., Williamson, L., Anderson, L., and Laman, J.D., "Reevaluating Biota Alteration: Reframing Environmental Influences on Chronic Immune Disorders and Exploring Novel Therapeutic Opportunities," *Yale Journal of Biology and Medicine* 97, no. 2 (2024): 253–63. https://doi.org/10.59249/vunf1315.

29 Smyth, K., Morton, C., Mathew, A., et al., "Production and Use of Hymenolepis Diminuta Cysticercoids as Anti-Inflammatory Therapeutics," *Journal of Clinical Medicine* 6, no. 10 (2017): 98. https://doi.org/10.3390/jcm6100098.

30 Bollinger, R.R., Everett, M.L., Palestrant, D., Love, S.D., Lin, S.S., and Parker, W., "Human Secretory Immunoglobulin a May Contribute to Biofilm Formation in the Gut," *Immunology* 109, no. 4 (2003): 580–87. https://doi.org/10.1046/j.1365-2567.2003.01700.x. Palestrant, D., Holzknecht, Z.E., Collins, B.H., Miller, S.E., Parker, W., and Bollinger, R.R., "Microbial Biofilms in the Gut: Visualization by Electron Microscopy and by Acridine Orange Staining," *Ultrastructural Pathology* 28 (2004): 23–27.

31 Parker, W., Jirků, K., Patel, E., Williamson, L., Anderson, L., and Laman, J.D., "Reevaluating Biota Alteration: Reframing Environmental Influences on Chronic Immune Disorders and Exploring Novel Therapeutic Opportunities," *Yale Journal of Biology and Medicine* 97, no. 2 (2024): 253–63. https://doi.org/10.59249/vunf1315.

32 Flynn, C.K., Adams, J.B., Krajmalnik-Brown, R., et al., "Review of Elevated Para-Cresol in Autism and Possible Impact on Symptoms," *International Journal of Molecular Sciences* 26, no. 4 (2025). https://doi.org/10.3390/ijms26041513. Xiong, X., Liu, D., Wang, Y., Zeng, T., and Peng, Y., "Urinary 3-(3-Hydroxyphenyl)-3-Hydroxypropionic Acid, 3-Hydroxyphenylacetic Acid, and 3-Hydroxyhippuric Acid Are Elevated in Children with Autism Spectrum Disorders," *BioMed Research International* 2016 (2016): 9485412. https://doi.org/10.1155/2016/9485412. Kang, D.-W., Adams, J.B., Gregory, A.C., et al., "Microbiota Transfer Therapy Alters Gut Ecosystem and Improves Gastrointestinal and Autism Symptoms: An

Open-Label Study," *Microbiome* 5, no. 1 (2017): 10. https://doi.org/10.1186/s40168-016-0225-7.

33 Schultz, S.T., Klonoff-Cohen, H.S., Wingard, D.L., Akshoomoff, N.A., Macera, C.A., and Ji, M., "Acetaminophen (Paracetamol) Use, Measles-Mumps-Rubella Vaccination, and Autistic Disorder. The Results of a Parent Survey," *Autism* 12, no. 3 (2008): 293–307.

34 Good, P., "Did Acetaminophen Provoke the Autism Epidemic?" *Alternative Medicine Review* 14, no. 4 (2009): 364–72.

35 Shaw, W., "Evidence That Increased Acetaminophen Use in Genetically Vulnerable Children Appears to Be a Major Cause of the Epidemics of Autism, Attention Deficit with Hyperactivity, and Asthma," *Journal of Restorative Medicine* 2 (2013): 1–16.

36 Frye, R.E., Slattery, J., MacFabe, D.F., et al. "Approaches to Studying and Manipulating the Enteric Microbiome to Improve Autism Symptoms," *Microbial Ecology in Health and Disease* 26 (2015): 26878. https://doi.org/10.3402/mehd.v26.26878.

37 Bilbo, S.D., Nevison, C.D., and Parker, W., "A Model for the Induction of Autism in the Ecosystem of the Human Body: The Anatomy of a Modern Pandemic?" *Microbial Ecology in Health and Disease* 26 (2015): 26253. https://doi.org/10.3402/mehd.v26.26253.

38 Bilbo, S.D., Nevison, C.D., and Parker, W., "A Model for the Induction of Autism in the Ecosystem of the Human Body: The Anatomy of a Modern Pandemic?" *Microbial Ecology in Health and Disease* 26 (2015): 26253. https://doi.org/10.3402/mehd.v26.26253.

39 Parker, W., Hornik, C.D., Bilbo, S., et al., "The Role of Oxidative Stress, Inflammation and Acetaminophen Exposure from Birth to Early Childhood in the Induction of Autism," *Journal of International Medical Research* 45, no. 2 (2017): 407–38.

40 Suda, N., Hernandez, J.C., Poulton, J., et al., "Therapeutic Doses of Paracetamol with Co-Administration of Cysteine and Mannitol During Early Development Result in Long Term Behavioral Changes in Laboratory Rats," *PLOS One* 16, no.6 (2020): e0253543. https://doi.org/https://doi.org/10.1371/journal.pone.0253543.

41 Ihrcke, N.S., Parker, W., Reissner, K.J., and Platt, J.L., "Regulation of Platelet Heparanase During Inflammation: Role of Ph and Proteinases," *Journal of Cellular Physiology* 175, no. 3 (1998): 255–67.
Parker, W., Holzknecht, Z.E., Song, A., et al., "Fate of Antigen in Xenotransplantation: Implications for Acute Vascular Rejection and Accommodation," *American Journal of Pathology* 152, no. 3 (1998): 829–39.

42 Williamson, L.L., McKenney, E.A., Holzknecht, Z.E., et al., "Got Worms? Perinatal Exposure to Helminths Prevents Persistent Immune Sensitization and Cognitive Dysfunction Induced by Early-Life Infection," *Brain, Behavior, and Immunity* 51 (2016): 14–28. https://doi.org/10.1016/j.bbi.2015.07.006.
McKenney, E.A., Williamson, L., Yoder, A.D., Rawls, J.F., Bilbo, S.D., and Parker, W., "Alteration of the Rat Cecal Microbiome During Colonization with

the Helminth *Hymenolepis Dimunita*," *Gut Microbes* 6 (2015): 182–93. https://
www.tandfonline.com/doi/full/10.1080/19490976.2015.1047128

43 Parker, W., Corrigan, P.T., Anderson, R., et al., "Evidence That Acetaminophen
Triggers Autism in Susceptible Individuals Has Been Ignored and Mishandled for
More Than a Decade," *Journal of the Academy of Public Health* (2025). https://
publichealth.realclearjournals.org/literature-syntheses/2025/10/evidence-that
-acetaminophen-triggers-autism-in-susceptible-individuals-has-been-ignored-and
-mishandled-for-more-than-a-decade/.
Patel, E., Jones,, J.P., 3rd, Bono-Lunn, D., et al., "The Safety of Pediatric Use
of Paracetamol (Acetaminophen): A Narrative Review of Direct and Indirect
Evidence," *Minerva Pediatrics (Torino)* 74, no. 6 (2022): 774–88. https://doi
.org/10.23736/s2724-5276.22.06932-4.
Zhao, L., Jones, J., Anderson, L., et al., "Acetaminophen Causes
Neurodevelopmental Injury in Susceptible Babies and Children: No Valid
Rationale for Controversy," *Clinical and Experimental Pediatrics* (2023). https://
doi.org/10.3345/cep.2022.01319.
Parker, W., Anderson, L.G., Jones, J.P., et al., "The Dangers of Acetaminophen
for Neurodevelopment Outweigh Scant Evidence for Long-Term Benefits,"
Children 11, no. 1 (2024): 44. https://www.mdpi.com/2227-9067/11/1/44.
Jones, J.P., 3rd, Williamson, L., Konsoula, Z., Anderson, R., Reissner, K.J., and
Parker, W., "Evaluating the Role of Susceptibility Inducing Cofactors and of
Acetaminophen in the Etiology of Autism Spectrum Disorder," *Life (Basel)* 14,
no. 8 (2024). https://doi.org/10.3390/life14080918.

44 Parker, W., Corrigan, P.T., Anderson, R., et al., "Evidence That Acetaminophen
Triggers Autism in Susceptible Individuals Has Been Ignored and Mishandled for
More Than a Decade," *Journal of the Academy of Public Health* (2025). https://
publichealth.realclearjournals.org/literature-syntheses/2025/10/evidence-that
-acetaminophen-triggers-autism-in-susceptible-individuals-has-been-ignored-and
-mishandled-for-more-than-a-decade/.

45 Zhao, L., Jones, J., Anderson, L., et al., "Acetaminophen Causes
Neurodevelopmental Injury in Susceptible Babies and Children: No Valid
Rationale for Controversy," *Clinical and Experimental Pediatrics* (2023). https://
doi.org/10.3345/cep.2022.01319.

46 Good, P., "Evidence the U.S. Autism Epidemic Initiated by Acetaminophen
(Tylenol) Is Aggravated by Oral Antibiotic Amoxicillin/Clavulanate (Augmentin)
and Now Exponentially by Herbicide Glyphosate (Roundup)," *Clinical Nutrition
ESPEN* 23 (2018): 171–83. https://doi.org/10.1016/j.clnesp.2017.10.005.
Shaw, W., "Hypothesis: 2 Major Environmental and Pharmaceutical Factors-
Acetaminophen Exposure and Gastrointestinal Overgrowth of Clostridia Bacteria
Induced by Ingestion of Glyphosate-Contaminated Foods-Dysregulate the
Developmental Protein Sonic Hedgehog and Are Major Causes of Autism,"
Integrative Medicine (Encinitas) 23, no. 3 (2024): 12–23.

Chapter 2: The History of Acetaminophen and Its Connection to Autism

47   Jaskch, V., "Phenacetin and Thallin for Children," *The Indian Medical Gazette* 24, no. 7 (1889): 215.

48   Zhao, L., Jones, J., Anderson, L., et al., "Acetaminophen Causes Neurodevelopmental Injury in Susceptible Babies and Children: No Valid Rationale for Controversy," *Clinical and Experimental Pediatrics* (2023). https://doi.org/10.3345/cep.2022.01319.

49   Sher, D.A., and Gibson, J.L., "Pioneering, Prodigious and Perspicacious: Grunya Efimovna Sukhareva's Life and Contribution to Conceptualising Autism and Schizophrenia," *European Child & Adolescent Psychiatry* 32, no. 3 (2023): 475–90. https://doi.org/10.1007/s00787-021-01875-7.

50   Kanner, L., "Autistic Disturbances of Affective Contact," *The Nervous Child* 2, no. 3 (1943): 217–50.

51   Asperger, H., "Die 'Autistischen Psychopathen' im Kindesalter," *Archiv* für *Psychiatrie und Nervenkrankheiten* 117, no. 1 (1944): 76–136. https://doi.org/10.1007/BF01837709.

52   Rimland, B., "The Autism Increase: Research Needed on the Vaccine Connection," *Autism Research Review International* 14, no. 1 (2000): 3.

53   Werner, E., and Dawson, G., "Validation of the Phenomenon of Autistic Regression Using Home Videotapes," *Archives of General Psychiatry* 62, no. 8 (2005): 889–95. https://doi.org/10.1001/archpsyc.62.8.889.

54   Rimland, B., "The Autism Increase: Research Needed on the Vaccine Connection," *Autism Research Review International* 14, no. 1 (2000): 3.

55   Sher, D.A., and Gibson, J.L., "Pioneering, Prodigious and Perspicacious: Grunya Efimovna Sukhareva's Life and Contribution to Conceptualising Autism and Schizophrenia," *European Child & Adolescent Psychiatry* 32, no. 3 (2023): 475–90. https://doi.org/10.1007/s00787-021-01875-7.

56   Zhao, L., Jones, J., Anderson, L., et al., "Acetaminophen Causes Neurodevelopmental Injury in Susceptible Babies and Children: No Valid Rationale for Controversy," *Clinical and Experimental Pediatrics* (2023). https://doi.org/10.3345/cep.2022.01319.
     ProPublica, "Use Only as Directed," Sept. 20, 2013, https://www.propublica.org/article/tylenol-mcneil-fda-use-only-as-directed.

57   Darwin, C., *The Descent of Man and Selection in Relation to Sex* (John Murray, 1871).
     Darwin, C., *On the Origin of Species by Means of Natural Selection, or the Preservation of Favoured Races in the Struggle for Life (1st Edition)* (John Murray, 1859).

58   Ji, Y., Azuine, R.E., Zhang, Y., et al., "Association of Cord Plasma Biomarkers of in Utero Acetaminophen Exposure with Risk of Attention-Deficit/Hyperactivity Disorder and Autism Spectrum Disorder in Childhood," *JAMA Psychiatry* 77, no. 2 (2020): 180–89. https://doi.org/10.1001/jamapsychiatry.2019.3259.

59   Frisch, M., and Simonsen, J., "Ritual Circumcision and Risk of Autism Spectrum Disorder in 0- to 9-Year-Old Boys: National Cohort Study in Denmark," *Journal of the Royal Society of Medicine* 108, no. 7 (2015): 266–79. https://doi.org/10.1177/0141076814565942.

60 Miani, A., Di Bernardo, G.A., Højgaard, A.D., et al., "Neonatal Male Circumcision Is Associated with Altered Adult Socio-Affective Processing," *Heliyon* 6, no. 11 (2020): e05566. https://doi.org/10.1016/j.heliyon.2020. e05566.

61 Matsuishi, T., Shiotsuki, Y., Yoshimura, K., Shoji, H., Imuta, F., and Yamashita,F., "High Prevalence of Infantile Autism in Kurume City, Japan," *Journal of Child Neurology* 2, no. 4 (1987): 268–71. https://doi.org/10.1177/ 088307388700200406.
Coury, D.L., and Nash, P.L., "Epidemiology and Etiology of Autistic Spectrum Disorders Difficult to Determine," *Pediatric Annals* 32, no. 10 (2003): 696–700. https://doi.org/10.3928/0090-4481-20031001-11.
Williams, J.G., Higgins, J.P., and Brayne, C.E., "Systematic Review of Prevalence Studies of Autism Spectrum Disorders," *Archives of Disease in Childhood* 91, no. 1 (2006): 8–15. https://doi.org/10.1136/adc.2004.062083.

62 Coury, D.L., and Nash, P.L., "Epidemiology and Etiology of Autistic Spectrum Disorders Difficult to Determine," *Pediatric Annals* 32, no. 10 (2003): 696–700. https://doi.org/10.3928/0090-4481-20031001-11.
Williams, J.G., Higgins, J.P., and Brayne, C.E., "Systematic Review of Prevalence Studies of Autism Spectrum Disorders," *Archives of Disease in Childhood* 91, no. 1 (2006): 8–15. https://doi.org/10.1136/adc.2004.062083.

63 Schultz, S.T., Klonoff-Cohen, H.S., Wingard, D.L., Akshoomoff, N.A., Macera, C.A., and Ji, M., "Acetaminophen (Paracetamol) Use, Measles-Mumps-Rubella Vaccination, and Autistic Disorder. The Results of a Parent Survey," *Autism* 12, no. 3 (2008): 293–307.

64 Maher, B., "Personal Genomes: The Case of the Missing Heritability," *Nature* 456, no. 7218 (2008): 18–21. https://doi.org/10.1038/456018a.

65 Ji, Y., Azuine, R.E., Zhang, Y., et al., "Association of Cord Plasma Biomarkers of in Utero Acetaminophen Exposure with Risk of Attention-Deficit/Hyperactivity Disorder and Autism Spectrum Disorder in Childhood," *JAMA Psychiatry* 77, no. 2 (2020): 180–89. https://doi.org/10.1001/jamapsychiatry.2019.3259.

66 Panghal, R., Mitra, S., Singh, J., Sarna, R., and Goel, B., "Oral Acetaminophen as an Adjunct to Continuous Epidural Infusion and Patient-Controlled Epidural Analgesia in Laboring Parturients: A Randomized Controlled Trial," *Journal of Anesthesia* 35, no. 6 (2021): 794–800. https://doi.org/10.1007/ s00540-021-02975-z.
Zutshi, V., Rani, K.U., Marwah, S., and Patel, M., "Efficacy of Intravenous Infusion of Acetaminophen for Intrapartum Analgesia," *Journal of Clinical and Diagnostic Research* 10, no. 8 (2016): Qc18–21. https://doi.org/10.7860/ jcdr/2016/19786.8375.

67 Schoenecker, J.G., Hauck, R.K., Mercer, M.C., Parker, W., and Lawson, J.H., "Exposure to Topical Bovine Thrombin During Surgery Elicits a Response against the Xenogeneic Carbohydrate Galactose(Alpha1-3)Galactose," *Journal of Clinical Immunology* 20, no. 6 (2000): 434–44.
Schoenecker, J.G., Johnson, R.K., Lesher, A.P., et al., "Exposure of Mice to Topical Bovine Thrombin Induces Systemic Autoimmunity," *American Journal of Pathology* 159, no. 5 (2001): 1957–69.

Schoenecker, J.G., Johnson, R.K., Fields, R.C., et al., "Relative Purity of Thrombin-Based Hemostatic Agents Used in Surgery," *Journal of the American College of Surgeons.* 197, no. 4 (2003): 580–90.
Lawson, J.H., Lynn, K.A., Domzalski, T., Ortel, T.A., Nicklason, L.E., and Parker, W., "Anti-Human Factor V Antibodies after Use of Relatively Pure Bovine Thrombin," *Annals of Thoracic Surgery* 75 (2005): 1037–38.

68 Parker, W., Corrigan, P.T., Anderson, R., et al., "Evidence That Acetaminophen Triggers Autism in Susceptible Individuals Has Been Ignored and Mishandled for More Than a Decade," *Journal of the Academy of Public Health* (2025). https://publichealth.realclearjournals.org/literature-syntheses/2025/10/evidence-that-acetaminophen-triggers-autism-in-susceptible-individuals-has-been-ignored-and-mishandled-for-more-than-a-decade/.

69 Zhao, L., Jones, J., Anderson, L., et al., "Acetaminophen Causes Neurodevelopmental Injury in Susceptible Babies and Children: No Valid Rationale for Controversy," *Clinical and Experimental Pediatrics* (2023). https://doi.org/10.3345/cep.2022.01319.

70 Zhao, L., Jones, J., Anderson, L., et al., "Acetaminophen Causes Neurodevelopmental Injury in Susceptible Babies and Children: No Valid Rationale for Controversy," *Clinical and Experimental Pediatrics* (2023). https://doi.org/10.3345/cep.2022.01319.

71 Good, P., "Did Acetaminophen Provoke the Autism Epidemic?" *Alternative Medicine Review* 14, no. 4 (2009): 364–72.

72 ProPublica, "Use Only as Directed," Sept. 20, 2013, https://www.propublica.org/article/tylenol-mcneil-fda-use-only-as-directed.

73 Prescott, L., "The Metabolism of Paracetamol," Chap. 6 In *Acetaminophen (Paracetamol) a Critical Bibliographic Review* (Taylor & Francis, 1996).

74 Alberti, A., Pirrone, P., Elia, M., Waring, R.H., and Romano, C., "Sulphation Deficit in 'Low-Functioning' Autistic Children: A Pilot Study," *Biological Psychiatry* 46, no. 3 (1999): 420–24. https://doi.org/10.1016/s0006-3223(98)00337-0.

75 Rimland, B., "The Autism Increase: Research Needed on the Vaccine Connection," *Autism Research Review International* 14, no. 1 (2000): 3.
Goldberg, D., "MMR, Autism, and Adam," *BMJ* 320, no. 7231 (2000): 38.

76 Parker, W., Hornik, C.D., Bilbo, S., et al., "The Role of Oxidative Stress, Inflammation and Acetaminophen Exposure from Birth to Early Childhood in the Induction of Autism," *Journal of International Medical Research* 45, no. 2 (2017): 407–38.

77 St Omer, V.V., and McKnight, E.D., 3rd., "Acetylcysteine for Treatment of Acetaminophen Toxicosis in the Cat," *Journal of the American Veterinary Medical Association* 176, no. 9 (1980): 911–13.

78 Court, M.H., "Feline Drug Metabolism and Disposition: Pharmacokinetic Evidence for Species Differences and Molecular Mechanisms," *Veterinary Clinics of North America: Small Animal Practices* 43, no. 5 (2013): 1039–54. https://doi.org/10.1016/j.cvsm.2013.05.002.

79 Green, M.D., Shires, T.K., and Fischer, L.J., "Hepatotoxicity of Acetaminophen in Neonatal and Young Rats. I. Age-Related Changes in Susceptibility,"

Toxicology and Applied Pharmacology 74, no. 1 (1984): 116–24. https://doi.org/10.1016/0041-008x(84)90277-1

80 Cendejas-Hernandez, J., Sarafian, J., Lawton, V., et al., "Paracetamol (Acetaminophen) Use in Infants and Children Was Never Shown to Be Safe for Neurodevelopment: A Systematic Review with Citation Tracking," *European Journal of Pediatrics* 181 (2022): 1835–57. https://doi.org/10.1007/s00431-022-04407-w.

81 Cendejas-Hernandez, J., Sarafian, J., Lawton, V., et al., "Paracetamol (Acetaminophen) Use in Infants and Children Was Never Shown to Be Safe for Neurodevelopment: A Systematic Review with Citation Tracking," *European Journal of Pediatrics* 181 (2022): 1835–57. https://doi.org/10.1007/s00431-022-04407-w.

82 Viberg, H., Eriksson, P., Gordh, T., and Fredriksson, A., "Paracetamol (Acetaminophen) Administration During Neonatal Brain Development Affects Cognitive Function and Alters Its Analgesic and Anxiolytic Response in Adult Male Mice," *Toxicological Sciences* 138, no. 1 (2013): 139–47. https://doi.org/10.1093/toxsci/kft329.

83 Zhao, L., Jones, J., Anderson, L., et al., "Acetaminophen Causes Neurodevelopmental Injury in Susceptible Babies and Children: No Valid Rationale for Controversy," *Clinical and Experimental Pediatrics* (2023). https://doi.org/10.3345/cep.2022.01319.

84 Patel, E., Jones, J.P., 3rd, Bono-Lunn, D., et al., "The Safety of Pediatric Use of Paracetamol (Acetaminophen): A Narrative Review of Direct and Indirect Evidence," *Minerva Pediatrics (Torino)* 74, no. 6 (2022): 774–88. https://doi.org/10.23736/s2724-5276.22.06932-4.

85 Cendejas-Hernandez, J., Sarafian, J., Lawton, V., et al., "Paracetamol (Acetaminophen) Use in Infants and Children Was Never Shown to Be Safe for Neurodevelopment: A Systematic Review with Citation Tracking," *European Journal of Pediatrics* 181 (2022): 1835–57. https://doi.org/10.1007/s00431-022-04407-w.

Chapter 3: The Single Most Blinding Scientific Error—Those Are Not Confounding Factors

86 Sheikh, J., Allotey, J., Sobhy, S., et al., "Maternal Paracetamol (Acetaminophen) Use During Pregnancy and Risk of Autism Spectrum Disorder and Attention Deficit/Hyperactivity Disorder in Offspring: Umbrella Review of Systematic Reviews," *BMJ* 391 (2025): e088141. https://doi.org/10.1136/bmj-2025-088141.

Editor, "Weaponizing Uncertainty in Science and in Public Health Puts People in Harm's Way," *Nature* 646, no. 8084 (2025): 259–60. https://doi.org/10.1038/d41586-025-03167-5.

Pearson, H., and Ledford, H., "Trump Links Autism and Tylenol: Is There Any Truth to It?" *Nature* 646, no. 8083 (2025): 13–14. https://doi.org/10.1038/d41586-025-02876-1.

Looi, M.K., and Bowie, K., "Autism: Trump Links Condition to Tylenol and Touts Leucovorin as 'First' US Therapeutic," *BMJ* 390 (2025): r2004. https://doi.org/10.1136/bmj.r2004.

Schweitzer, K., "Acetaminophen Use in Pregnancy-Study Author Explains the Data," *JAMA* 334, no. 17 (2025): 1499–501. https://doi.org/10.1001/jama.2025.19345.

Andrade, C., "Maternal Use of Acetaminophen (Paracetamol) During Pregnancy and Neurodevelopmental Disorders in Offspring: A Reasoned Evaluation of Risk," *Journal of Clinical Psychiatry* 86, no. 4 (2025). https://doi.org/10.4088/JCP.25f16187.

Louwen, F., Deuster, E., McAuliffe, F.M., et al., "Paracetamol (Acetaminophen) Use During Pregnancy and Autism Risk: Evidence Does Not Support Causal Association," *International Journal of Gynaecology & Obstetrics* 171, no. 3 (2025): 915–19. https://doi.org/10.1002/ijgo.70577.

Fombonne, E., "Editorial: The Acetaminophen Scare: Association vs Causation," *Journal of Child Psychology and Psychiatry* 66, no. 11 (2025): 1621–26. https://doi.org/10.1111/jcpp.70064.

Wise, J., "Paracetamol (Tylenol): No Clear Link between Use in Pregnancy and Autism or ADHD in Children, Rapid Review Finds," *BMJ* 391 (2025): r2368. https://doi.org/10.1136/bmj.r2368.

Lee, B.K., Stephansson, O., and Gardner, R.M., "Paracetamol (Acetaminophen) Use in Pregnancy and Risk of Autism and ADHD," *BMJ* 391 (2025): r2438. https://doi.org/10.1136/bmj.r2438.

Lang, K., "Trump's Claims on Tylenol (Paracetamol), Vaccines, and Autism—What's the Truth?" *BMJ* 390 (2025): r2025. https://doi.org/10.1136/bmj.r2025.

Aronson, J.K., "When I Use a Word . . . Paracetamol/Acetaminophen-Autism and Asthma," *BMJ* 390 (2025): r2032. https://doi.org/10.1136/bmj.r2032.

Gupta, S., Kopacz, K.S., and Hurdle, M.F.B., "Acetaminophen, Autism, and Analgesia in Pregnancy—A Pain Medicine Perspective," *Pain Management* 16, no. 1 (2026): 1–3. https://doi.org/10.1080/17581869.2025.2588240.

Ottewell, L., and Wright, F., "Purity, Politics and Pain: Trump's Paracetamol Posturing and the Moralisation of Pregnancy," *Medical Humanities* (2025). https://doi.org/10.1136/medhum-2025-013652.

Gardner, R.M., and Lee, B.K., "Triangulating Evidence on Prenatal Acetaminophen Use: Insights from a Large Japanese Cohort," *Paediatric and Perinatal Epidemiology* (2025). https://doi.org/10.1111/ppe.70105.

87   Editor, "Weaponizing Uncertainty in Science and in Public Health Puts People in Harm's Way," *Nature* 646, no. 8084 (2025): 259–60. https://doi.org/10.1038/d41586-025-03167-5.

Pearson, H., and Ledford, H., "Trump Links Autism and Tylenol: Is There Any Truth to It?" *Nature* 646, no. 8083 (2025): 13–14. https://doi.org/10.1038/d41586-025-02876-1.

88   Schweitzer, K., "Acetaminophen Use in Pregnancy-Study Author Explains the Data," *JAMA* 334, no. 17 (2025): 1499–501. https://doi.org/10.1001/jama.2025.19345.

89   Sheikh, J., Allotey, J., Sobhy, S., et al., "Maternal Paracetamol (Acetaminophen) Use During Pregnancy and Risk of Autism Spectrum Disorder and Attention Deficit/Hyperactivity Disorder in Offspring: Umbrella Review of Systematic Reviews," *BMJ* 391 (2025): e088141. https://doi.org/10.1136/bmj-2025-088141.
Looi, M.K., and Bowie, K., "Autism: Trump Links Condition to Tylenol and Touts Leucovorin as 'First' US Therapeutic," *BMJ* 390 (2025): r2004. https://doi.org/10.1136/bmj.r2004.
Wise, J., "Paracetamol (Tylenol): No Clear Link between Use in Pregnancy and Autism or ADHD in Children, Rapid Review Finds," *BMJ* 391 (2025): r2368. https://doi.org/10.1136/bmj.r2368.
Lee, B.K., Stephansson, O., and Gardner, R.M., "Paracetamol (Acetaminophen) Use in Pregnancy and Risk of Autism and ADHD," *BMJ* 391 (2025): r2438. https://doi.org/10.1136/bmj.r2438.
Lang, K., "Trump's Claims on Tylenol (Paracetamol), Vaccines, and Autism— What's the Truth?" *BMJ* 390 (2025): r2025. https://doi.org/10.1136/bmj.r2025.
Aronson, J.K., "When I Use a Word . . . Paracetamol/Acetaminophen-Autism and Asthma," *BMJ* 390 (2025): r2032. https://doi.org/10.1136/bmj.r2032.

90   Ahlqvist, V.H., Sjöqvist, H., Dalman, C., et al., "Acetaminophen Use During Pregnancy and Children's Risk of Autism, ADHD, and Intellectual Disability," *JAMA* 331, no. 14 (2024): 1205–14.

91   Whittemore, V., "Study Reveals No Causal Link between Neurodevelopmental Disorders and Acetaminophen Exposure before Birth," NIH news release, April 11, 2024.

92   Tyagi, U., and Barwal, K.C., "Ignaz Semmelweis—Father of Hand Hygiene," *Indian Journal of Surgery* 82, no. 3 (2020): 276–77. https://doi.org/10.1007/s12262-020-02386-6.
Semmelweis, I., *The Etiology, Concept and Prophylaxis of Childbed Fever, translated by K. Codell Carter*, edited by Coleman, W., D.C. Lindberg, and R.L. Numbers (The University of Wisconsin Press, 1983).

93   McFadden, J., "Razor Sharp: The Role of Occam's Razor in Science," *Annals of the New York Academy of Sciences* 1530, no. 1 (2023): 8–17. https://doi.org/https://doi.org/10.1111/nyas.15086.

94   Tang, N., Wu, Y., Ma, J., Wang, B., and Yu, R., "Coffee Consumption and Risk of Lung Cancer: A Meta-Analysis," *Lung Cancer* 67, no. 1 (2010): 17–22. https://doi.org/10.1016/j.lungcan.2009.03.012.

95   Kobylińska, Z., Biesiadecki, M., Kuna, E., Galiniak, S., and Mołoń, M., "Coffee as a Source of Antioxidants and an Elixir of Youth," *Antioxidants (Basel, Switzerland)* 14, no. 3 (2025). https://doi.org/10.3390/antiox14030285.

96   Guertin, K.A., Freedman, N.D., Loftfield, E., Graubard, B.I., Caporaso, N.E., and Sinha, R., "Coffee Consumption and Incidence of Lung Cancer in the NIH-AARP Diet and Health Study," *International Journal of Epidemiology* 45, no. 3 (2016): 929–39. https://doi.org/10.1093/ije/dyv104.

97   Parker, W., Hornik, C.D., Bilbo, S., et al., "The Role of Oxidative Stress, Inflammation and Acetaminophen Exposure from Birth to Early Childhood in

the Induction of Autism," *Journal of International Medical Research* 45, no. 2 (2017): 407–38.

98 Brandlistuen, R.E., Ystrom, E., Nulman, I., Koren, G., and Nordeng, H., "Prenatal Paracetamol Exposure and Child Neurodevelopment: A Sibling-Controlled Cohort Study," *International Journal of Epidemiology* 42, no. 6 (2013): 1702–13.

99 Parker, W., Hornik, C.D., Bilbo, S., et al., "The Role of Oxidative Stress, Inflammation and Acetaminophen Exposure from Birth to Early Childhood in the Induction of Autism," *Journal of International Medical Research* 45, no. 2 (2017): 407–38.

100 Biolo, G., Antonione, R., and De Cicco, M., "Glutathione Metabolism in Sepsis," *Critical Care Medicine* 35, no. 9 Suppl (2007): S591-95. https://doi .org/10.1097/01.ccm.0000278913.19123.13.

101 Bell, C.J.M., Mehta, M., Mirza, L., Young, A.H., and Beck, K., "Glutathione Alterations in Depression: A Meta-Analysis and Systematic Review of Proton Magnetic Resonance Spectroscopy Studies," *Psychopharmacology (Berl)* 242, no. 4 (2025): 717–24. https://doi.org/10.1007/s00213-024-06735-1.

102 Brandlistuen, R.E., Ystrom, E., Nulman, I., Koren, G., and Nordeng, H., "Prenatal Paracetamol Exposure and Child Neurodevelopment: A Sibling-Controlled Cohort Study," *International Journal of Epidemiology* 42, no. 6 (2013): 1702–13.

103 Parker, W., Hornik, C.D., Bilbo, S., et al., "The Role of Oxidative Stress, Inflammation and Acetaminophen Exposure from Birth to Early Childhood in the Induction of Autism," *Journal of International Medical Research* 45, no. 2 (2017): 407–38

104 Jones, J.P., 3rd, Williamson, L., Konsoula, Z., Anderson, R., Reissner, K.J., and Parker, W., "Evaluating the Role of Susceptibility Inducing Cofactors and of Acetaminophen in the Etiology of Autism Spectrum Disorder," *Life (Basel)* 14, no. 8 (2024).

105 Ahlqvist, V.H., Sjöqvist, H., Dalman, C., et al., "Acetaminophen Use During Pregnancy and Children's Risk of Autism, ADHD, and Intellectual Disability," *JAMA* 331, no. 14 (2024): 1205–14. https://doi.org/10.1001/jama.2024.3172.

106 Whittemore, V., "Study Reveals No Causal Link between Neurodevelopmental Disorders and Acetaminophen Exposure before Birth," NIH news release, April 11, 2024.

107 Ahlqvist, V.H., Sjöqvist, H., Dalman, C., et al., "Acetaminophen Use During Pregnancy and Children's Risk of Autism, ADHD, and Intellectual Disability," *JAMA* 331, no. 14 (2024): 1205–14. https://doi.org/10.1001/jama.2024.3172.

108 Jones, J.P., 3rd, Williamson, L., Konsoula, Z., Anderson, R., Reissner, K.J., and Parker, W., "Evaluating the Role of Susceptibility Inducing Cofactors and of Acetaminophen in the Etiology of Autism Spectrum Disorder," *Life (Basel)* 14, no. 8 (2024). https://doi.org/10.3390/life14080918.

109 Parker, W., Anderson, L.G., Jones, J.P., et al., "The Dangers of Acetaminophen for Neurodevelopment Outweigh Scant Evidence for Long-Term Benefits," *Children* 11, no. 1 (2024): 44. https://www.mdpi.com/2227-9067/11/1/44.

110 Jones, J.P., 3rd, Williamson, L., Konsoula, Z., Anderson, R., Reissner, K.J., and Parker, W., "Evaluating the Role of Susceptibility Inducing Cofactors and of Acetaminophen in the Etiology of Autism Spectrum Disorder," *Life (Basel)* 14, no. 8 (2024). https://doi.org/10.3390/life14080918.

111 Jones, J.P., 3rd, Williamson, L., Konsoula, Z., Anderson, R., Reissner, K.J., and Parker, W., "Evaluating the Role of Susceptibility Inducing Cofactors and of Acetaminophen in the Etiology of Autism Spectrum Disorder," *Life (Basel)* 14, no. 8 (2024). https://doi.org/10.3390/life14080918.

112 Jones, J.P., 3rd, Williamson, L., Konsoula, Z., Anderson, R., Reissner, K.J., and Parker, W., "Evaluating the Role of Susceptibility Inducing Cofactors and of Acetaminophen in the Etiology of Autism Spectrum Disorder," *Life (Basel)* 14, no. 8 (2024). https://doi.org/10.3390/life14080918.

113 Jones, J.P., 3rd, Williamson, L., Konsoula, Z., Anderson, R., Reissner, K.J., and Parker, W., "Evaluating the Role of Susceptibility Inducing Cofactors and of Acetaminophen in the Etiology of Autism Spectrum Disorder," *Life (Basel)* 14, no. 8 (2024). https://doi.org/10.3390/life14080918.

114 Sheikh, J., Allotey, J., Sobhy, S., et al., "Maternal Paracetamol (Acetaminophen) Use During Pregnancy and Risk of Autism Spectrum Disorder and Attention Deficit/Hyperactivity Disorder in Offspring: Umbrella Review of Systematic Reviews," *BMJ* 391 (2025): e088141. https://doi.org/10.1136/bmj -2025-088141.

115 Foo, J., Bellot, G., Pervaiz, S., and Alonso, S., "Mitochondria-Mediated Oxidative Stress During Viral Infection," *Trends in Microbiology* 30, no. 7 (2022): 679–92. https://doi.org/10.1016/j.tim.2021.12.011.
Milhelm, Z., Zanoaga, O., Pop, L., et al., "Evaluation of Oxidative Stress Biomarkers for Differentiating Bacterial and Viral Infections: A Comparative Study of Glutathione Disulfide (GSSG) and Reduced Glutathione (GSH)," *Medicine and Pharmacy Reports* 98, no. 1 (2025): 46–53. https://doi .org/10.15386/mpr-2821.
Shelton, R.C., Claiborne, J., Sidoryk-Wegrzynowicz, M., et al., "Altered Expression of Genes Involved in Inflammation and Apoptosis in Frontal Cortex in Major Depression," *Molecular Psychiatry* 16, no. 7 (2011): 751–62. https://doi .org/10.1038/mp.2010.52.
Berk, M., Williams, L.J., Jacka, F.N., et al., "So Depression Is an Inflammatory Disease, but Where Does the Inflammation Come From?" *BMC Medicine* 11 (2013): 200. https://doi.org/10.1186/1741-7015-11-200.
Salim, S., "Oxidative Stress and the Central Nervous System," *Journal of Pharmacology and Experimental Therapeutics* 360, no. 1 (2017): 201–5. https:// doi.org/10.1124/jpet.116.237503.
Lean, S.C., Jones, R.L., Roberts, S.A., and Heazell, A.E.P., "A Prospective Cohort Study Providing Insights for Markers of Adverse Pregnancy Outcome in Older Mothers," *BMC Pregnancy Childbirth* 21, no. 1 (2021): 706. https://doi .org/10.1186/s12884-021-04178-6.
Krishnamurthy, H.K., Rajavelu, I., Pereira, M., et al., "Inside the Genome: Understanding Genetic Influences on Oxidative Stress," *Frontiers in Genetics* 15 (2024): 1397352. https://doi.org/10.3389/fgene.2024.1397352.

Forsberg, L., de Faire, U., and Morgenstern, R., "Oxidative Stress, Human Genetic Variation, and Disease," *Archives of Biochemistry and Biophysics* 389, no. 1 (2001): 84–93. https://doi.org/10.1006/abbi.2001.2295.

Marseglia, L., Manti, S., D'Angelo, G., et al., "Oxidative Stress in Obesity: A Critical Component in Human Diseases," *International Journal of Molecular Sciences* 16, no. 1 (2014): 378–400. https://doi.org/10.3390/ijms16010378.

Moller, P., Wallin, H., and Knudsen, L.E., "Oxidative Stress Associated with Exercise, Psychological Stress and Life-Style Factors," *Chemico-Biological Interactions* 102, no. 1 (1996): 17–36. http://ukpmc.ac.uk/abstract/MED/8827060.

Kalghatgi, S., Spina, C.S., Costello, J.C., et al., "Bactericidal Antibiotics Induce Mitochondrial Dysfunction and Oxidative Damage in Mammalian Cells," *Science Translational Medicine* 5, no. 192 (2013): 192ra85. https://doi.org/10.1126/scitranslmed.3006055.

Skrabalova, J., Drastichova, Z., and Novotny, J., "Morphine as a Potential Oxidative Stress-Causing Agent," *Mini-Reviews in Organic Chemistry* 10, no. 4 (2013): 367–72. https://doi.org/10.2174/1570193x113106660031.

116 Palta, P., Szanton, S.L., Semba, R.D., Thorpe, R.J., Varadhan, R., and Fried, L.P., "Financial Strain Is Associated with Increased Oxidative Stress Levels: The Women's Health and Aging Studies," *Geriatric Nursing* 36, no. 2 Suppl (2015): S33–37. https://doi.org/10.1016/j.gerinurse.2015.02.020.

Eick, S.M., Barrett, E.S., van 't Erve, T.J., et al., "Association between Prenatal Psychological Stress and Oxidative Stress During Pregnancy," *Paediatric and Perinatal Epidemiology* 32, no. 4 (2018): 318–26. https://doi.org/10.1111/ppe.12465.

117 Parker, W., Hornik, C.D., Bilbo, S., et al., "The Role of Oxidative Stress, Inflammation and Acetaminophen Exposure from Birth to Early Childhood in the Induction of Autism," *Journal of International Medical Research* 45, no. 2 (2017): 407–38.

118 Patel, E., Jones, J.P., 3rd, Bono-Lunn, D., et al., "The Safety of Pediatric Use of Paracetamol (Acetaminophen): A Narrative Review of Direct and Indirect Evidence," *Minerva Pediatrics (Torino)* 74, no. 6 (2022): 774–88. https://doi.org/10.23736/s2724-5276.22.06932-4.

119 Ahlqvist, V.H., Sjöqvist, H., Dalman, C., et al., "Acetaminophen Use During Pregnancy and Children's Risk of Autism, ADHD, and Intellectual Disability," *JAMA* 331, no. 14 (2024): 1205–14. https://doi.org/10.1001/jama.2024.3172.

120 Jones, J.P., 3rd, Williamson, L., Konsoula, Z., Anderson, R., Reissner, K.J., and Parker, W., "Evaluating the Role of Susceptibility Inducing Cofactors and of Acetaminophen in the Etiology of Autism Spectrum Disorder," *Life (Basel)* 14, no. 8 (2024). https://doi.org/10.3390/life14080918.

121 Parker, W., Hornik, C.D., Bilbo, S., et al., "The Role of Oxidative Stress, Inflammation and Acetaminophen Exposure from Birth to Early Childhood in the Induction of Autism," *Journal of International Medical Research* 45, no. 2 (2017): 407–38.

Parker, W., Anderson, L.G., Jones, J.P., et al., "The Dangers of Acetaminophen for Neurodevelopment Outweigh Scant Evidence for Long-Term Benefits," *Children* 11, no. 1 (2024): 44. https://www.mdpi.com/2227-9067/11/1/44. Jones, J.P., 3rd, Williamson, L., Konsoula, Z., Anderson, R., Reissner, K.J., and Parker, W., "Evaluating the Role of Susceptibility Inducing Cofactors and of Acetaminophen in the Etiology of Autism Spectrum Disorder," *Life (Basel)* 14, no. 8 (2024). https://doi.org/10.3390/life14080918.

122 Yousif, S.R., Aboody, R., and Keil, F.C., "The Illusion of Consensus: A Failure to Distinguish between True and False Consensus," *Psychological Science* 30, no. 8 (2019): 1195–204. https://doi.org/10.1177/0956797619856844.

123 Prada, D., Ritz, B., Bauer, A.Z., and Baccarelli, A.A., "Evaluation of the Evidence on Acetaminophen Use and Neurodevelopmental Disorders Using the Navigation Guide Methodology," *Environmental Health* 24, no. 1 (2025): 56. https://doi.org/10.1186/s12940-025-01208-0.

124 Prada, D., Ritz, B., Bauer, A.Z., and Baccarelli, A.A., "Evaluation of the Evidence on Acetaminophen Use and Neurodevelopmental Disorders Using the Navigation Guide Methodology," *Environmental Health* 24, no. 1 (2025): 56. https://doi.org/10.1186/s12940-025-01208-0.

125 Bollinger, R.B., Barbas, A.S., Bush, E.L., Lin, S.S., and Parker, W., "Biofilms in the Normal Human Large Bowel: Fact Rather Than Fiction," *Gut* 56 (2007): 1481–82.

126 Etzel, R.A., Grandjean, P., and Ozonoff, D.M., "Environmental Epidemiology in a Crossfire," *Environmental Health* 20, no. 1 (2021): 91. https://doi.org/10.1186/s12940-021-00776-1.

127 Etzel, R.A., Grandjean, P., and Ozonoff, D.M., "Environmental Epidemiology in a Crossfire," *Environmental Health* 20, no. 1 (2021): 91. https://doi.org/10.1186/s12940-021-00776-1.

Chapter 4: Associations Between Acetaminophen and Autism in Place and Time

128 Kim, H.U., "Autism across Cultures: Rethinking Autism," *Disability & Society* 27, no. 4 (2012): 535–45. https://doi.org/10.1080/09687599.2012.659463.

129 Spinazzi, N.A., Santoro, J.D., Pawlowski, K., et al., "Co-Occurring Conditions in Children with Down Syndrome and Autism: A Retrospective Study," *Journal of Neurodevelopmental Disorders* 15, no. 1 (2023): 9. https://doi.org/10.1186/s11689-023-09478-w.

130 Maenner, M.J., Warren, Z., Williams, A.R., et al., "Prevalence and Characteristics of Autism Spectrum Disorder among Children Aged 8 Years—Autism and Developmental Disabilities Monitoring Network, 11 Sites, United States, 2020," *MMWR Surveillance Summaries* 72, no. 2 (2023): 1–14. https://doi.org/10.15585/mmwr.ss7202a1.

131 Neph, A., "A Comparison of Healthcare between Nicaragua and the United States and the Feasibility of Naturopathic Medicine," *Honors Projects* 73 (2011). https://scholarworks.gvsu.edu/honorsprojects/73.

Vargas-Palacios, E., Pineda, R., and Galán-Rodas, E., "The Politicised and Crumbling Nicaraguan Health System," *Lancet* 392, no. 10165 (2018): 2694–95. https://doi.org/10.1016/s0140-6736(18)32990-8.

132 Lotter, V., "Childhood Autism in Africa," *Journal of Child Psychology and Psychiatry* 19, no. 3 (1978): 231–44. https://doi.org/10.1111/j.1469-7610.1978.tb00466.x.

133 Lotter, V., "Childhood Autism in Africa," *Journal of Child Psychology and Psychiatry* 19, no. 3 (1978): 231–44. https://doi.org/10.1111/j.1469-7610.1978.tb00466.x.

134 Decoteau, C.L., "The 'Western Disease': Autism and Somali Parents' Embodied Health Movements," *Social Science & Medicine* 177 (2017): 169–76. https://doi.org/10.1016/j.socscimed.2017.01.064.

135 Aguiar, A.G., Mainegra, F.D., García, R.O., and Hernandez, F.Y., "Diagnosis in Children with Autism Spectrum Disorders in Their Development in Textual Comprehension," *Revista de Ciencias Medicas de Pinar del Río* 20, no. 6 (2016): 729–37.

136 Pharmacy Times, "A Peek at Pharmacy Practice in Cuba," May 24, 2016, https://www.pharmacytimes.com/view/a-peek-at-pharmacy-practice-in-cuba.

137 C. Yeldham, "Going to Cuba? Here's What Else to Pack," STARTUP CUBA.TV., Nov. 10, 2021, https://startupcuba.tv/2021/11/10/going-to-cuba-heres-what-else-to-pack/.
B. Sainsbury, "20 Things to Know before Visiting Cuba," Lonely Planet, Dec. 23, 2025, https://www.lonelyplanet.com/articles/things-to-know-before-traveling-to-cuba.
Simply Cuba Tours, "Preparing for Cuba: The Ultimate Packing List for Health & Safety," Dec. 11, 2025, https://simplycubatours.com/preparing-for-cuba-the-ultimate-packing-list-for-health-safety/.
Locally Sourced Cuba Tours, "How to Support Local Cubans When You Travel to Cuba," accessed Feb. 7, 2026, https://locallysourcedcuba.com/how-to-support-locals-in-cuba/.

138 Kamer, A., Zohar, A.H., Youngmann, R., Diamond, G.W., Inbar, D., and Senecky, Y., "A Prevalence Estimate of Pervasive Developmental Disorder among Immigrants to Israel and Israeli Natives A File Review Study," *Social Psychiatry and Psychiatric Epidemiology* 39, no. 2 (2004): 141–45. https://doi.org/10.1007/s00127-004-0696-x.

139 American Academy of Pediatrics, "Acetaminophen Is Safe for Children When Taken as Directed, No Link to Autism," updated Oct. 29, 2025. https://www.aap.org/en/news-room/fact-checked/acetaminophen-is-safe-for-children-when-taken-as-directed-no-link-to-autism/.

140 Robinson, J.L., Nations, L., Suslowitz, N., Cuccaro, M.L., Haines, J., and Pericak-Vance, M., *Prevalence Rates of Autism Spectrum Disorders among the Old Order Amish* (International Society for Autism Research, 2010).

141 Parker, W., Anderson, L.G., Jones, J.P., et al., "The Dangers of Acetaminophen for Neurodevelopment Outweigh Scant Evidence for Long-Term Benefits," *Children* 11, no. 1 (2024): 44. https://www.mdpi.com/2227-9067/11/1/44.

142 Parker, W., Anderson, L.G., Jones, J.P., et al., "The Dangers of Acetaminophen for Neurodevelopment Outweigh Scant Evidence for Long-Term Benefits," *Children* 11, no. 1 (2024): 44. https://www.mdpi.com/2227-9067/11/1/44.

143 Parker, W., Anderson, L.G., Jones, J.P., et al., "The Dangers of Acetaminophen for Neurodevelopment Outweigh Scant Evidence for Long-Term Benefits," *Children* 11, no. 1 (2024): 44. https://www.mdpi.com/2227-9067/11/1/44.

144 Jones, J., Konsoula, Z., Williamson, L., Anderson, R., Meza-Keuthen, S., and Parker, W., "Three Mandatory Doses of Acetaminophen During the First Months of Life with the MenB Vaccine: A Protocol for the Induction of Autism Spectrum Disorder in Susceptible Individuals," Preprints, 2025. https://doi.org/10.20944/preprints202501.0319.v5.
Dungan, A.R., VanRyzin, J.W., Wood, J.A., et al., "Early Life Exposure to Acetaminophen Exerts Cohort-Dependent Effects on Nursing and Play Behavior in Sprague Dawley Rats," *SSRN (Preprint)* http://dx.doi.org/10.2139/ssrn.5223387 (2025). http://dx.doi.org/10.2139/ssrn.5223387.

145 Parker, W., Anderson, L.G., Jones, J.P., et al., "The Dangers of Acetaminophen for Neurodevelopment Outweigh Scant Evidence for Long-Term Benefits," *Children* 11, no. 1 (2024): 44. https://www.mdpi.com/2227-9067/11/1/44.

146 Parker, W., Anderson, L.G., Jones, J.P., et al., "The Dangers of Acetaminophen for Neurodevelopment Outweigh Scant Evidence for Long-Term Benefits," *Children* 11, no. 1 (2024): 44. https://www.mdpi.com/2227-9067/11/1/44.

147 Jones, J.P., 3rd, Williamson, L., Konsoula, Z., Anderson, R., Reissner, K.J., and Parker, W., "Evaluating the Role of Susceptibility Inducing Cofactors and of Acetaminophen in the Etiology of Autism Spectrum Disorder," *Life (Basel)* 14, no. 8 (2024). https://doi.org/10.3390/life14080918

148 Easton, T., and Herrera, S., "J&J's Dirty Little Secret," *Forbes,* Jan. 12, 1998, https://www.forbes.com/forbes/1998/0112/6101042a.html.
ProPublica, "Use Only as Directed," Sept. 20, 2013, https://www.propublica.org/article/tylenol-mcneil-fda-use-only-as-directed.

149 Nevison, C., Blaxill, M., and Zahorodny, W., "California Autism Prevalence Trends from 1931 to 2014 and Comparison to National ASD Data from Idea and ADDM," *Journal of Autism and Developmental Disorders* 48, no. 12 (2018): 4103–17. https://doi.org/10.1007/s10803-018-3670-2.

150 Rimland, B., "The Autism Increase: Research Needed on the Vaccine Connection," *Autism Research Review International* 14, no. 1 (2000): 3.

151 Zhao, L., Jones, J., Anderson, L., et al., "Acetaminophen Causes Neurodevelopmental Injury in Susceptible Babies and Children: No Valid Rationale for Controversy," *Clinical and Experimental Pediatrics* (2023). https://doi.org/10.3345/cep.2022.01319.

152 Zhao, L., Jones, J., Anderson, L., et al., "Acetaminophen Causes Neurodevelopmental Injury in Susceptible Babies and Children: No Valid Rationale for Controversy," *Clinical and Experimental Pediatrics* (2023). https://doi.org/10.3345/cep.2022.01319.

153 Wisk, L.E., and Sharma, N., "Prevalence and Trends in Pediatric-Onset Chronic Conditions in the United States, 1999–2018," *Academic Pediatrics* 25, no. 4 (2025): 102810. https://doi.org/10.1016/j.acap.2025.102810.

154 Matsuishi, T., Shiotsuki, Y., Yoshimura, K., Shoji, H., Imuta, F., and Yamashita, F., "High Prevalence of Infantile Autism in Kurume City, Japan," *Journal of Child Neurology* 2, no. 4 (1987): 268–71. https://doi.org/10.1177/088307388700200406.

155 Williams, J.G., Higgins, J.P., and Brayne, C.E., "Systematic Review of Prevalence Studies of Autism Spectrum Disorders," *Archives of Disease in Childhood* 91, no. 1 (2006): 8–15. https://doi.org/10.1136/adc.2004.062083.

156 Coury, D.L., and Nash, P.L., "Epidemiology and Etiology of Autistic Spectrum Disorders Difficult to Determine," *Pediatric Annals* 32, no. 10 (2003): 696–700. https://doi.org/10.3928/0090-4481-20031001-11.

157 Dungan, A.R., VanRyzin, J.W., Wood, J.A., et al., "Early Life Exposure to Acetaminophen Exerts Cohort-Dependent Effects on Nursing and Play Behavior in Sprague Dawley Rats," *SSRN (Preprint)* http://dx.doi.org/10.2139/ssrn.5223387 (2025).

Chapter 5: Timing of Risk: Pregnancy, Labor and Delivery, and Early Childhood

158 Parker, W., Anderson, L.G., Jones, J.P., et al., "The Dangers of Acetaminophen for Neurodevelopment Outweigh Scant Evidence for Long-Term Benefits," *Children* 11, no. 1 (2024): 44. https://www.mdpi.com/2227-9067/11/1/44.

159 Miners, J.O., Robson, R.A., and Birkett, D.J., "Paracetamol Metabolism in Pregnancy," *British Journal of Clinical Pharmacology* 22, no. 3 (1986): 359–62. https://doi.org/10.1111/j.1365-2125.1986.tb02901.x.

160 Zuppa, A.F., Hammer, G.B., Barrett, J.S., et al., "Safety and Population Pharmacokinetic Analysis of Intravenous Acetaminophen in Neonates, Infants, Children, and Adolescents with Pain or Fever," *The Journal of Pediatric Pharmacology and Therapeutics: JPPT: The Official Journal of PPAG* 16, no. 4 (2011): 246–61. https://doi.org/10.5863/1551-6776-16.4.246.

161 Zuppa, A.F., Hammer, G.B., Barrett, J.S., et al., "Safety and Population Pharmacokinetic Analysis of Intravenous Acetaminophen in Neonates, Infants, Children, and Adolescents with Pain or Fever," *The Journal of Pediatric Pharmacology and Therapeutics: JPPT: The Official Journal of PPAG* 16, no. 4 (2011): 246–61. https://doi.org/10.5863/1551-6776-16.4.246.

162 Alberti, A., Pirrone, P., Elia, M., Waring, R.H., and Romano, C., "Sulphation Deficit in 'Low-Functioning' Autistic Children: A Pilot Study," *Biological Psychiatry* 46, no. 3 (1999): 420–24. https://doi.org/10.1016/s0006-3223(98)00337-0.
Geier, D.A., Kern, J.K., Garver, C.R., Adams, J.B., Audhya, T., and Geier, M.R., "A Prospective Study of Transsulfuration Biomarkers in Autistic Disorders," *Neurochemical Research* 34, no. 2 (2009): 386–93. https://doi.org/10.1007/s11064-008-9782-x.
Pagan, C., Benabou, M., Leblond, C., et al., "Decreased Phenol Sulfotransferase Activities Associated with Hyperserotonemia in Autism Spectrum Disorders," *Translational Psychiatry* 11, no. 1 (2021): 23. https://doi.org/10.1038/s41398-020-01125-5.

163 Stein, T.P., Schluter, M.D., Steer, R.A., and Ming, X., "Autism and Phthalate Metabolite Glucuronidation," *Journal of Autism and Developmental Disorders* 43, no. 11 (2013): 2677–85. https://doi.org/10.1007/s10803-013-1822-y.

164 Ji, Y., Azuine, R.E., Zhang, Y., et al., "Association of Cord Plasma Biomarkers of in Utero Acetaminophen Exposure with Risk of Attention-Deficit/Hyperactivity Disorder and Autism Spectrum Disorder in Childhood," *JAMA Psychiatry* 77, no. 2 (2020): 180–89. https://doi.org/10.1001/jamapsychiatry.2019.3259.

165 Parker, W., Anderson, L.G., Jones, J.P., et al., "The Dangers of Acetaminophen for Neurodevelopment Outweigh Scant Evidence for Long-Term Benefits," *Children* 11, no. 1 (2024): 44. https://www.mdpi.com/2227-9067/11/1/44.

166 Tan, C., Frewer, V., Cox, G., Williams, K., and Ure, A., "Prevalence and Age of Onset of Regression in Children with Autism Spectrum Disorder: A Systematic Review and Meta-Analytical Update," *Autism Research* 14, no. 3 (2021): 582–98.

167 Tan, C., Frewer, V., Cox, G., Williams, K., and Ure, A., "Prevalence and Age of Onset of Regression in Children with Autism Spectrum Disorder: A Systematic Review and Meta-Analytical Update," *Autism Research* 14, no. 3 (2021): 582–98.

168 Ahlqvist, V.H., Sjöqvist, H., Dalman, C., et al., "Acetaminophen Use During Pregnancy and Children's Risk of Autism, Adhd, and Intellectual Disability," *JAMA* 331, no. 14 (2024): 1205–14. https://doi.org/10.1001/jama.2024.3172.

169 Jones, J.P., 3rd, Williamson, L., Konsoula, Z., Anderson, R., Reissner, K.J., and Parker, W., "Evaluating the Role of Susceptibility Inducing Cofactors and of Acetaminophen in the Etiology of Autism Spectrum Disorder," *Life (Basel)* 14, no. 8 (2024). https://doi.org/10.3390/life14080918.

170 Jones, J.P., 3rd, Williamson, L., Konsoula, Z., Anderson, R., Reissner, K.J., and Parker, W., "Evaluating the Role of Susceptibility Inducing Cofactors and of Acetaminophen in the Etiology of Autism Spectrum Disorder," *Life (Basel)* 14, no. 8 (2024). https://doi.org/10.3390/life14080918.
Bornehag, C.G., Reichenberg, A., Hallerback, M.U., et al., "Prenatal Exposure to Acetaminophen and Children's Language Development at 30 Months," *European Psychiatry* 51 (2018): 98–103. https://doi.org/10.1016/j.eurpsy.2017.10.007.

171 Jones, J.P., 3rd, Williamson, L., Konsoula, Z., Anderson, R., Reissner, K.J., and Parker, W., "Evaluating the Role of Susceptibility Inducing Cofactors and of Acetaminophen in the Etiology of Autism Spectrum Disorder," *Life (Basel)* 14, no. 8 (2024). https://doi.org/10.3390/life14080918.

172 Murray, C.B., de la Vega, R., Murphy, L.K., Kashikar-Zuck, S., and Palermo, T.M, "The Prevalence of Chronic Pain in Young Adults: A Systematic Review and Meta-analysis," *Pain* 163 (2022): e972–e984.

173 Patel, E., Jones J.P., 3rd, Bono-Lunn, D., et al., "The Safety of Pediatric Use of Paracetamol (Acetaminophen): A Narrative Review of Direct and Indirect Evidence," *Minerva Pediatrics (Torino)* 74, no. 6 (2022): 774–88. https://doi.org/10.23736/s2724-5276.22.06932-4.

174 Jones, J.P., 3rd, Williamson, L., Konsoula, Z., Anderson, R., Reissner, K.J., and Parker, W., "Evaluating the Role of Susceptibility Inducing Cofactors and of Acetaminophen in the Etiology of Autism Spectrum Disorder," *Life (Basel)* 14, no. 8 (2024). https://doi.org/10.3390/life14080918.

175 Zhao, L., Jones, J., Anderson, L., et al., "Acetaminophen Causes Neurodevelopmental Injury in Susceptible Babies and Children: No Valid Rationale for Controversy," *Clinical and Experimental Pediatrics* (2023). https://doi.org/10.3345/cep.2022.01319.

176 Werner, E., and Dawson, G., "Validation of the Phenomenon of Autistic Regression Using Home Videotapes," *Archives of General Psychiatry* 62, no. 8 (2005): 889–95. https://doi.org/10.1001/archpsyc.62.8.889.

177 Zhao, L., Jones, J., Anderson, L., et al., "Acetaminophen Causes Neurodevelopmental Injury in Susceptible Babies and Children: No Valid Rationale for Controversy," *Clinical and Experimental Pediatrics* (2023). https://doi.org/10.3345/cep.2022.01319.

178 Freed, G.L., Clark, S.J., Butchart, A.T., Singer, D.C., and Davis, M.M., "Parental Vaccine Safety Concerns in 2009," *Pediatrics* 125, no. 4 (2010): 654–59. https://doi.org/10.1542/peds.2009-1962.
Bazzano, A., Zeldin, A., Schuster, E., Barrett, C., and Lehrer, D., "Vaccine-Related Beliefs and Practices of Parents of Children with Autism Spectrum Disorders," *American Journal of Intellectual and Developmental Disabilities* 117, no. 3 (2012): 233–42. https://doi.org/10.1352/1944-7558-117.3.233.

179 Wakefield, A.J., Murch, S.H., Anthony, A., et al., "Ileal-Lymphoid-Nodular Hyperplasia, Non-Specific Colitis, and Pervasive Developmental Disorder in Children," *Lancet* 351, no. 9103 (1998): 637–41. https://doi.org/10.1016/s0140-6736(97)11096-0.

Chapter 6: Laboratory Animal Studies and Evident Toxicity

180 Viberg, H., Eriksson, P., Gordh, T., and Fredriksson, A., "Paracetamol (Acetaminophen) Administration During Neonatal Brain Development Affects Cognitive Function and Alters Its Analgesic and Anxiolytic Response in Adult Male Mice," *Toxicological Sciences* 138, no. 1 (2013): 139–47. https://doi.org/10.1093/toxsci/kft329.

181 Patel, E., Jones, J.P., 3rd, Bono-Lunn, D., et al., "The Safety of Pediatric Use of Paracetamol (Acetaminophen): A Narrative Review of Direct and Indirect Evidence," *Minerva Pediatrics (Torino)* 74, no. 6 (2022): 774–88. https://doi.org/10.23736/s2724-5276.22.06932-4.

182 Suda, N., Hernandez, J.C., Poulton, J., et al., "Therapeutic Doses of Paracetamol with Co-Administration of Cysteine and Mannitol During Early Development Result in Long Term Behavioral Changes in Laboratory Rats," *PLOS One* 16, no. 6 (2020): e0253543. https://doi.org/https://doi.org/10.1371/journal.pone.0253543.

183 Suda, N., Hernandez, J.C., Poulton, J., et al., "Therapeutic Doses of Paracetamol with Co-Administration of Cysteine and Mannitol During Early Development Result in Long Term Behavioral Changes in Laboratory Rats," *PLOS One* 16, no. 6 (2020): e0253543. https://doi.org/https://doi.org/10.1371/journal.pone.0253543.

184 Dean, S.L., Knutson, J.F., Krebs-Kraft, D.L., and McCarthy, M.M., "Prostaglandin E2 Is an Endogenous Modulator of Cerebellar Development and Complex Behavior During a Sensitive Postnatal Period," *European Journal*

of Neuroscience 35, no. 8 (2012): 1218–29. https://doi.org/10.1111/
j.1460-9568.2012.08032.x.

185 Herrington, J.A., Guss Darwich, J., Harshaw, C., Brigande, A.M., Leif, E.B.,
and Currie, P.J., "Elevated Ghrelin Alters the Behavioral Effects of Perinatal
Acetaminophen Exposure in Rats," *Developmental Psychobiology* 64, no. 3 (2022):
e22252. https://doi.org/10.1002/dev.22252.

186 Harshaw, C., and Warner, A.G., "Interleukin-1β-Induced Inflammation
and Acetaminophen During Infancy: Distinct and Interactive Effects on
Social-Emotional and Repetitive Behavior in C57BL/6J Mice," *Pharmacology
Biochemistry and Behavior* 220 (2022): 173463. https://doi.org/https://doi
.org/10.1016/j.pbb.2022.173463.

187 Hussin, S., and Al-Allaf, L., "Histological Changes of Ca and Dg Regions of
Hippocampus of Rats' Brain after Exposure to Acetaminophen in Postnatal
Period," *Iraqi Journal of Veterinary Sciences* 36, no. 1 (2022): 151–58.

188 Klein, R.M., Motomura, V.N., Debiasi, J.D., and Moreira, E.G., "Gestational
Paracetamol Exposure Induces Core Behaviors of Neurodevelopmental Disorders
in Infant Rats and Modifies Response to a Cannabinoid Agonist in Females,"
Neurotoxicology and Teratology 99 (2023): 107279. https://doi.org/10.1016/
j.ntt.2023.107279.
Klein, R.M., Rigobello, C., Vidigal, C.B., et al., "Gestational Exposure to
Paracetamol in Rats Induces Neurofunctional Alterations in the Progeny,"
Neurotoxicology and Teratology 77 (2020): 106838. https://doi.org/10.1016/
j.ntt.2019.106838.

189 Graeca, M., and Kulesza, R., "Impaired Brainstem Auditory Evoked Potentials
after in Utero Exposure to High Dose Paracetamol Exposure," *Hearing Research*
454 (2024): 109149. https://doi.org/10.1016/j.heares.2024.109149.

190 Saad, A., Hegde, S., Kechichian, T., et al., "Is There a Causal Relation between
Maternal Acetaminophen Administration and ADHD?" *PLOS One* 11, no. 6
(2016): e0157380. https://doi.org/10.1371/journal.pone.0157380.

191 Saad, A., Hegde, S., Kechichian, T., et al., "Is There a Causal Relation between
Maternal Acetaminophen Administration and ADHD?" *PLOS One* 11, no. 6
(2016): e0157380. https://doi.org/10.1371/journal.pone.0157380.

192 Baker, B.H., Rafikian, E.E., Hamblin, P.B., Strait, M.D., Yang, M., and Pearson,
B.L., "Sex-Specific Neurobehavioral and Prefrontal Cortex Gene Expression
Alterations Following Developmental Acetaminophen Exposure in Mice,"
Neurobiology of Disease 177 (2023): 105970. https://doi.org/10.1016/
j.nbd.2022.105970.

193 Hay-Schmidt, A., Finkielman, O.T.E., Jensen, B.A.H., et al., "Prenatal Exposure
to Paracetamol/Acetaminophen and Precursor Aniline Impairs Masculinisation of
Male Brain and Behaviour," *Reproduction* 154, no. 2 (2017): 145–52. https://doi
.org/10.1530/rep-17-0165.

194 Blecharz-Klin, K., Wawer, A., Jawna-Zboińska, K., et al., "Early Paracetamol
Exposure Decreases Brain-Derived Neurotrophic Factor (Bdnf) in Striatum and
Affects Social Behaviour and Exploration in Rats," *Pharmacology Biochemistry and
Behavior* 168 (2018): 25–32. https://doi.org/10.1016/j.pbb.2018.03.004.

195 Baker, B.H., Rafikian, E.E., Hamblin, P.B., Strait, M.D., Yang, M., and Pearson, B.L., "Sex-Specific Neurobehavioral and Prefrontal Cortex Gene Expression Alterations Following Developmental Acetaminophen Exposure in Mice," *Neurobiology of Disease* 177 (2023): 105970. https://doi.org/10.1016/j.nbd.2022.105970.
Dean, S.L., Knutson, J.F., Krebs-Kraft, D.L., and McCarthy, M.M., "Prostaglandin E2 Is an Endogenous Modulator of Cerebellar Development and Complex Behavior During a Sensitive Postnatal Period," *European Journal of Neuroscience* 35, no. 8 (2012): 1218–29. https://doi.org/10.1111/j.1460-9568.2012.08032.x.
Kanno, S.I., Tomizawa, A., Yomogida, S., and Hara, A., "Glutathione Peroxidase 3 Is a Protective Factor against Acetaminophen-Induced Hepatotoxicity in Vivo and in Vitro," *International Journal of Molecular Medicine* 40, no. 3 (2017): 748–54. https://doi.org/10.3892/ijmm.2017.3049.
Rigobello, C., Klein, R.M., Debiasi, J.D., et al., "Perinatal Exposure to Paracetamol: Dose and Sex-Dependent Effects in Behaviour and Brain's Oxidative Stress Markers in Progeny," *Behavioural Brain Research* 408 (2021): 113294. https://doi.org/10.1016/j.bbr.2021.113294.

196 Graeca, M., and Kulesza, R., "Impaired Brainstem Auditory Evoked Potentials after in Utero Exposure to High Dose Paracetamol Exposure," *Hearing Research* 454 (2024): 109149. https://doi.org/10.1016/j.heares.2024.109149.

197 Klein, R.M., Rigobello, C., Vidigal, C.B., et al., "Gestational Exposure to Paracetamol in Rats Induces Neurofunctional Alterations in the Progeny," *Neurotoxicology and Teratology* 77 (2020): 106838. https://doi.org/10.1016/j.ntt.2019.106838.

198 Graeca, M., and Kulesza, R., "Impaired Brainstem Auditory Evoked Potentials after in Utero Exposure to High Dose Paracetamol Exposure," *Hearing Research* 454 (2024): 109149. https://doi.org/10.1016/j.heares.2024.109149.

199 Xu, M., Minagawa, Y., Kumazaki, H., Okada, K.I., and Naoi, N., "Prefrontal Responses to Odors in Individuals with Autism Spectrum Disorders: Functional Nirs Measurement Combined with a Fragrance Pulse Ejection System," *Frontiers in Human Neuroscience* 14 (2020): 523456. https://doi.org/10.3389/fnhum.2020.523456.

200 Chopyk, D.M., Stuart, J.D., Zimmerman, M.G., et al., "Acetaminophen Intoxication Rapidly Induces Apoptosis of Intestinal Crypt Stem Cells and Enhances Intestinal Permeability," *Hepatology Communications* 3, no. 11 (2019): 1435–49. https://doi.org/10.1002/hep4.1406.
Schneider, K.M., Elfers, C., Ghallab, A., et al., "Intestinal Dysbiosis Amplifies Acetaminophen-Induced Acute Liver Injury," *Cellular and Molecular Gastroenterology and Hepatology* 11, no. 4 (2021): 909–33.
Alabbas, S.Y., Giri, R., Oancea, I., et al., "Gut Inflammation and Adaptive Immunity Amplify Acetaminophen Toxicity in Bowel and Liver," *Journal of Gastroenterology and Hepatology* 38, no. 4 (2023): 609–18. https://doi.org/https://doi.org/10.1111/jgh.16102.

201 Fattorusso, A., Di Genova, L., Dell'Isola, G.B., Mencaroni, E., and Esposito, S., "Autism Spectrum Disorders and the Gut Microbiota," *Nutrients* 11, no. 3 (2019). https://doi.org/10.3390/nu11030521.
Peralta-Marzal, L.N., Prince, N., Bajic, D., et al., "The Impact of Gut Microbiota-Derived Metabolites in Autism Spectrum Disorders," *International Journal of Molecular Sciences* 22, no. 18 (2021). https://doi.org/10.3390/ijms221810052.

202 Flynn, C.K., Adams, J.B., Krajmalnik-Brown, R., et al., "Review of Elevated Para-Cresol in Autism and Possible Impact on Symptoms," *International Journal of Molecular Sciences* 26, no. 4 (2025). https://doi.org/10.3390/ijms26041513.
Xiong, X., Liu, D., Wang, Y., Zeng, T., and Peng, Y., "Urinary 3-(3-Hydroxyphenyl)-3-Hydroxypropionic Acid, 3-Hydroxyphenylacetic Acid, and 3-Hydroxyhippuric Acid Are Elevated in Children with Autism Spectrum Disorders," *BioMed Research International* 2016 (2016): 9485412. https://doi.org/10.1155/2016/9485412.

203 Clayton, T.A., Baker, D., Lindon, J.C., Everett, J.R., and Nicholson, J.K., "Pharmacometabonomic Identification of a Significant Host-Microbiome Metabolic Interaction Affecting Human Drug Metabolism," *Proceedings of the National Academy of Sciences of the United States of America* 106, no. 34 (2009): 14728–33. https://doi.org/10.1073/pnas.0904489106.

204 Close, R., Kremer, S., Mitchem, M., et al., "A Sulfotransferase from a Gut Microbe Acts on Diverse Phenolic Sulfate Compounds, Including Acetaminophen Sulfate," *PNAS Nexus* (2025). https://doi.org/10.1093/pnasnexus/pgaf403.

205 Flynn, C.K., Adams, J.B., Krajmalnik-Brown, R., et al., "Review of Elevated Para-Cresol in Autism and Possible Impact on Symptoms," *International Journal of Molecular Sciences* 26, no. 4 (2025). https://doi.org/10.3390/ijms26041513.
Xiong, X., Liu, D., Wang, Y., Zeng, T., and Peng, Y., "Urinary 3-(3-Hydroxyphenyl)-3-Hydroxypropionic Acid, 3-Hydroxyphenylacetic Acid, and 3-Hydroxyhippuric Acid Are Elevated in Children with Autism Spectrum Disorders," *BioMed Research International* 2016 (2016): 9485412. https://doi.org/10.1155/2016/9485412.

206 Dungan, A.R., VanRyzin, J.W., Wood, J.A., et al., "Early Life Exposure to Acetaminophen Exerts Cohort-Dependent Effects on Nursing and Play Behavior in Sprague Dawley Rats," *SSRN (Preprint)* http://dx.doi.org/10.2139/ssrn.5223387 (2025).

207 Dungan, A.R., VanRyzin, J.W., Wood, J.A., et al., "Early Life Exposure to Acetaminophen Exerts Cohort-Dependent Effects on Nursing and Play Behavior in Sprague Dawley Rats," *SSRN (Preprint)* http://dx.doi.org/10.2139/ssrn.5223387 (2025).

208 Suda, N., Hernandez, J.C., Poulton, J., et al., "Therapeutic Doses of Paracetamol with Co-Administration of Cysteine and Mannitol During Early Development Result in Long Term Behavioral Changes in Laboratory Rats," *PLOS One* 16, no. 6 (2020): e0253543. https://doi.org/https://doi.org/10.1371/journal.pone.0253543.

209 Saad, A., Hegde, S., Kechichian, T., et al., "Is There a Causal Relation between Maternal Acetaminophen Administration and ADHD?" *PLOS One* 11, no. 6 (2016): e0157380. https://doi.org/10.1371/journal.pone.0157380.

210 Auerbach, J., "Does New York City Really Have as Many Rats as People?" *Significance* 11, no. 4 (2014): 22–27. https://doi.org/https://doi.org/10.1111/j.1740-9713.2014.00764.x.

Chapter 7: More Evidence, from the Amazing to the Obscure

211 St Omer, V.V., and McKnight, E.D., 3rd., "Acetylcysteine for Treatment of Acetaminophen Toxicosis in the Cat," *Journal of the American Veterinary Medical Association* 176, no. 9 (1980): 911–13.
Court, M.H., "Feline Drug Metabolism and Disposition: Pharmacokinetic Evidence for Species Differences and Molecular Mechanisms," *Veterinary Clinics of North America: Small Animal Practices* 43, no. 5 (2013): 1039–54. https://doi.org/10.1016/j.cvsm.2013.05.002.
Anvik, J.O., "Acetaminophen Toxicosis in a Cat," *Canadian Veterinary Journal* 25, no. 12 (1984): 445–47.
Savides, M.C., Oehme, F.W., Nash, S.L., and Leipold, H.W., "The Toxicity and Biotransformation of Single Doses of Acetaminophen in Dogs and Cats," *Toxicology and Applied Pharmacology* 74, no. 1 (1984): 26–34. https://doi.org/10.1016/0041-008x(84)90266-7.
Lautz, L.S., Jeddi, M.Z., Girolami, F., Nebbia, C., and Dorne, J., "Metabolism and Pharmacokinetics of Pharmaceuticals in Cats (Felix Sylvestris Catus) and Implications for the Risk Assessment of Feed Additives and Contaminants," *Toxicology Letters* 338 (2021): 114–27. https://doi.org/10.1016/j.toxlet.2020.11.014.

212 Miller, R.P., Roberts, R.J., and Fischer, L.J., "Acetaminophen Elimination Kinetics in Neonates, Children, and Adults," *Clinical Pharmacology & Therapeutics* 19, no. 3 (1976): 284–94. https://doi.org/10.1002/cpt1976193284.
Cook, S.F., Stockmann, C., Samiee-Zafarghandy, S., et al., "Neonatal Maturation of Paracetamol (Acetaminophen) Glucuronidation, Sulfation, and Oxidation Based on a Parent-Metabolite Population Pharmacokinetic Model," *Clinical Pharmacokinetics* 55, no. 11 (2016): 1395–411. https://doi.org/10.1007/s40262-016-0408-1.

213 Levy, G., Khanna, N.N., Soda, D.M., Tsuzuki, O., and Stern, L., "Pharmacokinetics of Acetaminophen in the Human Neonate: Formation of Acetaminophen Glucuronide and Sulfate in Relation to Plasma Bilirubin Concentration and D-Glucaric Acid Excretion," *Pediatrics* 55, no. 6 (1975): 818–25.

214 Cook, S.F., Stockmann, C., Samiee-Zafarghandy, S., et al., "Neonatal Maturation of Paracetamol (Acetaminophen) Glucuronidation, Sulfation, and Oxidation Based on a Parent-Metabolite Population Pharmacokinetic Model," *Clinical Pharmacokinetics* 55, no. 11 (2016): 1395–411. https://doi.org/10.1007/s40262-016-0408-1.

215 Mazaleuskaya, L.L., Sangkuhl, K., Thorn, C.F., FitzGerald, G.A., Altman, R.B., and Klein, T.E., "PharmGKB Summary: Pathways of Acetaminophen Metabolism at the Therapeutic Versus Toxic Doses," *Pharmacogenet Genomics* 25, no. 8 (2015): 416–26. https://doi.org/10.1097/fpc.0000000000000150.

216 Williams, R.J., "Sulfate Deficiency as a Risk Factor for Autism," *Journal of Autism and Developmental Disorders* 50, no. 1 (2020): 153–61. https://doi.org/10.1007/s10803-019-04240-5.

217 Williams, R.J., "Sulfate Deficiency as a Risk Factor for Autism," *Journal of Autism and Developmental Disorders* 50, no. 1 (2020): 153–61. https://doi.org/10.1007/s10803-019-04240-5.

218 Alberti, A., Pirrone, P., Elia, M., Waring, R.H., and Romano, C., "Sulphation Deficit in 'Low-Functioning' Autistic Children: A Pilot Study," *Biological Psychiatry* 46, no. 3 (1999): 420–24. https://doi.org/10.1016/s0006-3223(98)00337-0.

219 Geier, D.A., Kern, J.K., Garver, C.R., Adams, J.B., Audhya, T., and Geier, M.R., "A Prospective Study of Transsulfuration Biomarkers in Autistic Disorders," *Neurochemical Research* 34, no. 2 (2009): 386–93. https://doi.org/10.1007/s11064-008-9782-x.
Pagan, C., Benabou, M., Leblond, C., et al., "Decreased Phenol Sulfotransferase Activities Associated with Hyperserotonemia in Autism Spectrum Disorders," *Translational Psychiatry* 11, no. 1 (2021): 23. https://doi.org/10.1038/s41398-020-01125-5.

220 Parker, W., Corrigan, P.T., Anderson, R., et al., "Evidence That Acetaminophen Triggers Autism in Susceptible Individuals Has Been Ignored and Mishandled for More Than a Decade," *Journal of the Academy of Public Health* (2025). https://publichealth.realclearjournals.org/literature-syntheses/2025/10/evidence-that-acetaminophen-triggers-autism-in-susceptible-individuals-has-been-ignored-and-mishandled-for-more-than-a-decade/.

221 Schultz, S.T., Klonoff-Cohen, H.S., Wingard, D.L., Akshoomoff, N.A., Macera, C.A., and Ji, M., "Acetaminophen (Paracetamol) Use, Measles-Mumps-Rubella Vaccination, and Autistic Disorder. The Results of a Parent Survey," *Autism* 12, no. 3 (2008): 293–307.

222 Parker, W., Corrigan, P.T., Anderson, R., et al., "Evidence That Acetaminophen Triggers Autism in Susceptible Individuals Has Been Ignored and Mishandled for More Than a Decade," *Journal of the Academy of Public Health* (2025). https://publichealth.realclearjournals.org/literature-syntheses/2025/10/evidence-that-acetaminophen-triggers-autism-in-susceptible-individuals-has-been-ignored-and-mishandled-for-more-than-a-decade/.

223 Frisch, M., and Simonsen, J., "Ritual Circumcision and Risk of Autism Spectrum Disorder in 0- to 9-Year-Old Boys: National Cohort Study in Denmark," *Journal of the Royal Society of Medicine* 108, no. 7 (2015): 266–79. https://doi.org/10.1177/0141076814565942.

224 Parker, W., Corrigan, P.T., Anderson, R., et al., "Evidence That Acetaminophen Triggers Autism in Susceptible Individuals Has Been Ignored and Mishandled for More Than a Decade," *Journal of the Academy of Public Health* (2025). https://publichealth.realclearjournals.org/literature-syntheses/2025/10/

evidence-that-acetaminophen-triggers-autism-in-susceptible-individuals-has
-been-ignored-and-mishandled-for-more-than-a-decade/.

225 Parker, W., Corrigan, P.T., Anderson, R., et al., "Evidence That Acetaminophen
Triggers Autism in Susceptible Individuals Has Been Ignored and Mishandled for
More Than a Decade," *Journal of the Academy of Public Health* (2025). https://
publichealth.realclearjournals.org/literature-syntheses/2025/10/evidence-that
-acetaminophen-triggers-autism-in-susceptible-individuals-has-been-ignored-and
-mishandled-for-more-than-a-decade/.

226 Morris, B.J., Moreton, S., Krieger, J.N., and Klausner, J.D., "Infant Circumcision
for Sexually Transmitted Infection Risk Reduction Globally," *Global Health
Science and Practice* 10, no. 4 (2022). https://doi.org/10.9745/ghsp-d-21-00811.
Morris, B.J., Katelaris, A., Blumenthal, N.J., et al., "Evidence-Based
Circumcision Policy for Australia," *Journal of Men's Health* 18, no. 6 (2022).
https://doi.org/10.31083/j.jomh1806132.
Morris, B.J., Moreton, S., Bailis, S.A., Cox, G., and Krieger, J.N., "Critical
Evaluation of Contrasting Evidence on Whether Male Circumcision Has Adverse
Psychological Effects: A Systematic Review," *Journal of Evidence-Based Medicine*
15, no. 2 (2022): 123–35. https://doi.org/10.1111/jebm.12482.
Morris, B.J., Moreton, S., and Krieger, J.N., "Critical Evaluation of Arguments
Opposing Male Circumcision: A Systematic Review," *Journal of Evidence-Based
Medicine* 12, no. 4 (2019): 263–90. https://doi.org/10.1111/jebm.12361.
Morris, B.J., Krieger, J.N., and Klausner, J.D., "CDC's Male Circumcision
Recommendations Represent a Key Public Health Measure,' *Global Health
Science and Practice* 5, no. 1 (2017): 15–27. https://doi.org/10.9745/
ghsp-d-16-00390.
Morris, B.J., and Wiswell, T.E., "'Circumcision Pain' Unlikely to Cause Autism,"
Journal of the Royal Society of Medicine 108, no. 8 (2015): 297.

227 Parker, W., Corrigan, P.T., Anderson, R., et al., "Evidence That Acetaminophen
Triggers Autism in Susceptible Individuals Has Been Ignored and Mishandled for
More Than a Decade," *Journal of the Academy of Public Health* (2025). https://
publichealth.realclearjournals.org/literature-syntheses/2025/10/evidence-that
-acetaminophen-triggers-autism-in-susceptible-individuals-has-been-ignored-and
-mishandled-for-more-than-a-decade/.

228 Yoon, E., Babar, A., Choudhary, M., Kutner, M., and Pyrsopoulos, N.,
"Acetaminophen-Induced Hepatotoxicity: A Comprehensive Update," *Journal
of Clinical and Translational Hepatology* 4, no. 2 (2016): 131–42. https://doi
.org/10.14218/jcth.2015.00052.

229 Green, M.D., Shires, T.K., and Fischer, L.J., "Hepatotoxicity of Acetaminophen
in Neonatal and Young Rats. I. Age-Related Changes in Susceptibility,"
Toxicology and Applied Pharmacology 74, no. 1 (1984): 116–24. https://doi
.org/10.1016/0041-008x(84)90277-1.

230 Rumack, B.H., "Aspirin Versus Acetaminophen: A Comparative View," *Pediatrics*
62, no. 5 Pt 2 Suppl (1978): 943–46.

231 Juujärvi, S., Saarela, T., Pokka, T., Hallman, M., and Aikio, O., "Intravenous
Paracetamol for Neonates: Long-Term Diseases Not Escalated During 5 Years

of Follow-Up," *Archives of Disease in Childhood: Fetal and Neonatal Edition* 106, no. 2 (2021): 178–83. https://doi.org/10.1136/archdischild-2020-319069. Juujärvi, S., Saarela, T., Hallman, M., and Aikio, O., "Trial of Paracetamol for Premature Newborns: Five-Year Follow-Up," *Journal of Maternal-Fetal & Neonatal Medicine* (2021): 1–3. https://doi.org/10.1080/14767058.2021.1875444.

232 Cendejas-Hernandez, J., Sarafian, J., Lawton, V., et al., "Paracetamol (Acetaminophen) Use in Infants and Children Was Never Shown to Be Safe for Neurodevelopment: A Systematic Review with Citation Tracking," *European Journal of Pediatrics* 181 (2022): 1835–57. https://doi.org/10.1007/s00431-022-04407-w.

233 DeWall, C.N., MacDonald, G., Webster, G.D., et al., "Acetaminophen Reduces Social Pain: Behavioral and Neural Evidence," *Psychological Science* 21, no. 7 (2010): 931–37. https://doi.org/10.1177/0956797610374741.

234 Roberts, I.D., Krajbich, I., and Way, B.M., "Acetaminophen Influences Social and Economic Trust," *Scientific Reports* 9, no. 1 (2019): 4060. https://doi.org/10.1038/s41598-019-40093-9.

235 Randles, D., Heine, S.J., and Santos, N., "The Common Pain of Surrealism and Death: Acetaminophen Reduces Compensatory Affirmation Following Meaning Threats," *Psychological Science* 24, no. 6 (2013): 966–73. https://doi.org/10.1177/0956797612464786.

236 Durso, G.R.O., Luttrell, A., and Way, B.M., "Over-the-Counter Relief from Pains and Pleasures Alike: Acetaminophen Blunts Evaluation Sensitivity to Both Negative and Positive Stimuli," *Psychological Science* 26, no. 6 (2015): 750–58. https://doi.org/10.1177/0956797615570366.

237 Mischkowski, D., Crocker, J., and Way, B.M., "A Social Analgesic? Acetaminophen (Paracetamol) Reduces Positive Empathy," *Frontiers in Psychology* 10 (2019): 538. https://doi.org/10.3389/fpsyg.2019.00538. Mischkowski, D., Crocker, J., and Way, B.M., "From Painkiller to Empathy Killer: Acetaminophen (Paracetamol) Reduces Empathy for Pain," *Social Cognitive and Affective Neuroscience* 11, no. 9 (2016): 1345–53. https://doi.org/10.1093/scan/nsw057

238 Chopyk, D.M., Stuart, J.D., Zimmerman, M.G., et al., "Acetaminophen Intoxication Rapidly Induces Apoptosis of Intestinal Crypt Stem Cells and Enhances Intestinal Permeability," *Hepatology Communications* 3, no. 11 (2019): 1435–49. https://doi.org/10.1002/hep4.1406. Schneider, K.M., Elfers, C., Ghallab, A., et al., "Intestinal Dysbiosis Amplifies Acetaminophen-Induced Acute Liver Injury," *Cellular and Molecular Gastroenterology and Hepatology* 11, no. 4 (2021): 909–33. Alabbas, S.Y., Giri, R., Oancea, I., et al., "Gut Inflammation and Adaptive Immunity Amplify Acetaminophen Toxicity in Bowel and Liver," *Journal of Gastroenterology and Hepatology* 38, no. 4 (2023): 609–18. https://doi.org/https://doi.org/10.1111/jgh.16102. McCrae, J.C., Morrison, E.E., MacIntyre, I.M., Dear, J.W., and Webb, D.J., "Long-Term Adverse Effects of Paracetamol—A Review," *British Journal of*

         Clinical Pharmacology 84, no. 10 (2018): 2218–30. https://doi.org/10.1111/
         bcp.13656.
239 Flynn, C.K., Adams, J.B., Krajmalnik-Brown, R., et al., "Review of Elevated
         Para-Cresol in Autism and Possible Impact on Symptoms," *International Journal
         of Molecular Sciences* 26, no. 4 (2025). https://doi.org/10.3390/ijms26041513.
         Xiong, X., Liu, D., Wang, Y., Zeng, T., and Peng, Y., "Urinary
         3-(3-Hydroxyphenyl)-3-Hydroxypropionic Acid, 3-Hydroxyphenylacetic Acid,
         and 3-Hydroxyhippuric Acid Are Elevated in Children with Autism Spectrum
         Disorders," *BioMed Research International* 2016 (2016): 9485412. https://doi
         .org/10.1155/2016/9485412.
         Fattorusso, A., Di Genova, L., Dell'Isola, G.B., Mencaroni, E., and Esposito,
         S., "Autism Spectrum Disorders and the Gut Microbiota," *Nutrients* 11, no. 3
         (2019). https://doi.org/10.3390/nu11030521.
         Peralta-Marzal, L.N., Prince, N., Bajic, D., et al., "The Impact of Gut
         Microbiota-Derived Metabolites in Autism Spectrum Disorders," *International
         Journal of Molecular Sciences* 22, no. 18 (2021). https://doi.org/10.3390/
         ijms221810052.
240 Flynn, C.K., Adams, J.B., Krajmalnik-Brown, R., et al., "Review of Elevated
         Para-Cresol in Autism and Possible Impact on Symptoms," *International Journal
         of Molecular Sciences* 26, no. 4 (2025). https://doi.org/10.3390/ijms26041513.
         Xiong, X., Liu, D., Wang, Y., Zeng, T., and Peng, Y., "Urinary
         3-(3-Hydroxyphenyl)-3-Hydroxypropionic Acid, 3-Hydroxyphenylacetic Acid,
         and 3-Hydroxyhippuric Acid Are Elevated in Children with Autism Spectrum
         Disorders," *BioMed Research International* 2016 (2016): 9485412. https://doi
         .org/10.1155/2016/9485412.
241 Hutabarat, R.M., Unadkat, J.D., Kushmerick, P., Aitken, M.L., Slattery, J.T.,
         and Smith, A.L., "Disposition of Drugs in Cystic Fibrosis. III. Acetaminophen,"
         Clinical Pharmacology & Therapeutics 50, no. 6 (1991): 695–701.
         Kearns, G.L., "Hepatic Drug Metabolism in Cystic Fibrosis: Recent
         Developments and Future Directions," *Annals of Pharmacotherapy* 27, no. 1
         (1993): 74–79.
242 Parker, W., Hornik, C.D., Bilbo, S., et al., "The Role of Oxidative Stress,
         Inflammation and Acetaminophen Exposure from Birth to Early Childhood in
         the Induction of Autism," *Journal of International Medical Research* 45, no. 2
         (2017): 407–38.
243 Parker, W., Hornik, C.D., Bilbo, S., et al., "The Role of Oxidative Stress,
         Inflammation and Acetaminophen Exposure from Birth to Early Childhood in
         the Induction of Autism," *Journal of International Medical Research* 45, no. 2
         (2017): 407–38.
244 Jones, J.P., 3rd, Williamson, L., Konsoula, Z., Anderson, R., Reissner, K.J., and
         Parker, W., "Evaluating the Role of Susceptibility Inducing Cofactors and of
         Acetaminophen in the Etiology of Autism Spectrum Disorder," *Life (Basel)* 14,
         no. 8 (2024). https://doi.org/10.3390/life14080918.
245 Reid, N., Shanley, D.C., Logan, J., White, C., Liu, W., and Hawkins, E.,
         "International Survey of Specialist Fetal Alcohol Spectrum Disorder Diagnostic
         Clinics: Comparison of Diagnostic Approach and Considerations Regarding the

Potential for Unification," *International Journal of Environmental Research and Public Health* 19, no. 23 (2022). https://doi.org/10.3390/ijerph192315663.
Hus, Y., and Segal, O., "Challenges Surrounding the Diagnosis of Autism in Children," *Neuropsychiatric Disease and Treatment* 17 (2021): 3509–29. https://doi.org/10.2147/ndt.s282569.

246 Hen-Herbst, L., Jirikowic, T., Hsu, L.Y., and McCoy, S.W., "Motor Performance and Sensory Processing Behaviors among Children with Fetal Alcohol Spectrum Disorders Compared to Children with Developmental Coordination Disorders," *Research in Developmental Disabilities* 103 (2020): 103680. https://doi.org/10.1016/j.ridd.2020.103680.
Coll, S.-M., Foster, N.E.V., Meilleur, A., Brambati, S.M., and Hyde, K.L., "Sensorimotor Skills in Autism Spectrum Disorder: A Meta-Analysis," *Research in Autism Spectrum Disorders* 76 (2020): 101570. https://doi.org/https://doi.org/10.1016/j.rasd.2020.101570.

247 Antshel, K.M., and Russo, N., "Autism Spectrum Disorders and ADHD: Overlapping Phenomenology, Diagnostic Issues, and Treatment Considerations," *Current Psychiatry Reports* 21, no. 5 (2019): 34. https://doi.org/10.1007/s11920-019-1020-5.
Rasmussen, C., Benz, J., Pei, J., et al., "The Impact of an ADHD Co-Morbidity on the Diagnosis of FASD," *Canadian Journal of Clinical Pharmacology* 17, no. 1 (2010): e165–76.

248 Lai, M.C., Kassee, C., Besney, R., et al., "Prevalence of Co-Occurring Mental Health Diagnoses in the Autism Population: A Systematic Review and Meta-Analysis," *Lancet Psychiatry* 6, no. 10 (2019): 819–29. https://doi.org/10.1016/s2215-0366(19)30289-5.
Weyrauch, D., Schwartz, M., Hart, B., Klug, M.G., and Burd, L., "Comorbid Mental Disorders in Fetal Alcohol Spectrum Disorders: A Systematic Review," *Journal of Developmental & Behavioral Pediatrics* 38, no. 4 (2017): 283–91. https://doi.org/10.1097/dbp.0000000000000440.

249 Matson, J.L., and Shoemaker, M., "Intellectual Disability and Its Relationship to Autism Spectrum Disorders," *Research in Developmental Disabilities* 30, no. 6 (2009): 1107–14. https://doi.org/10.1016/j.ridd.2009.06.003.
Chokroborty-Hoque, A., Alberry, B., and Singh, S.M., "Exploring the Complexity of Intellectual Disability in Fetal Alcohol Spectrum Disorders," *Frontiers in Pediatrics* 2 (2014): 90. https://doi.org/10.3389/fped.2014.00090.

250 Reid, N., Moritz, K.M., and Akison, L.K., "Adverse Health Outcomes Associated with Fetal Alcohol Exposure: A Systematic Review Focused on Immune-Related Outcomes," *Pediatric Allergy and Immunology* 30, no. 7 (2019): 698–707. https://doi.org/10.1111/pai.13099.
Amos-Kroohs, R.M., Fink, B.A., Smith, C.J., et al., "Abnormal Eating Behaviors Are Common in Children with Fetal Alcohol Spectrum Disorder," *Journal of Pediatrics* 169 (2016): 194–200.e1. https://doi.org/10.1016/j.jpeds.2015.10.049.
Hanlon-Dearman, A., Chen, M.L., and Olson, H.C., "Understanding and Managing Sleep Disruption in Children with Fetal Alcohol Spectrum Disorder," *Biochemistry and Cell Biology* 96, no. 2 (2018): 267–74. https://doi.org/10.1139/bcb-2017-0064.

Bell, S.H., Stade, B., Reynolds, J.N., et al., "The Remarkably High Prevalence of Epilepsy and Seizure History in Fetal Alcohol Spectrum Disorders," *Alcoholism Clinical and Experimental Research* 34, no. 6 (2010): 1084–89. https://doi .org/10.1111/j.1530-0277.2010.01184.x.
Al-Beltagi, M., "Autism Medical Comorbidities," *World Journal of Clinical Pediatrics* 10, no. 3 (2021): 15–28. https://doi.org/10.5409/wjcp.v10.i3.15.

251 Wei, H., Zhu, Y., Wang, T., Zhang, X., Zhang, K., and Zhang, Z., "Genetic Risk Factors for Autism-Spectrum Disorders: A Systematic Review Based on Systematic Reviews and Meta-Analysis," *Journal of Neural Transmission (Vienna)* 128, no. 6 (2021): 717–34.
Sambo, D., and Goldman, D., "Genetic Influences on Fetal Alcohol Spectrum Disorder," *Genes (Basel)* 14, no. 1 (2023). https://doi.org/10.3390/ genes14010195.

252 Modabbernia, A., Velthorst, E., and Reichenberg, A., "Environmental Risk Factors for Autism: An Evidence-Based Review of Systematic Reviews and Meta-Analyses," *Molecular Autism* 8 (2017): 13. https://doi.org/10.1186/ s13229-017-0121-4.
May, P.A., and Gossage, J.P., "Maternal Risk Factors for Fetal Alcohol Spectrum Disorders: Not as Simple as It Might Seem," *Alcohol Research & Health* 34, no. 1 (2011): 15–26.

253 Thompson, T., Oram, C., Correll, C.U., Tsermentseli, S., and Stubbs, B., "Analgesic Effects of Alcohol: A Systematic Review and Meta-Analysis of Controlled Experimental Studies in Healthy Participants," *The Journal of Pain* 18, no. 5 (2017): 499–510. https://doi.org/10.1016/j.jpain.2016.11.009.
Ohashi, N., and Kohno, T., "Analgesic Effect of Acetaminophen: A Review of Known and Novel Mechanisms of Action," *Frontiers in Pharmacology* 11 (2020): 580289. https://doi.org/10.3389/fphar.2020.580289.
Cruz, J.V., Maba, I.K., Correia, D., Kaziuk, F.D., Cadena, S., and Zampronio, A.R., "Intermittent Binge-Like Ethanol Exposure During Adolescence Attenuates the Febrile Response by Reducing Brown Adipose Tissue Thermogenesis in Rats," *Drug and Alcohol Dependence* 209 (2020): 107904. https://doi.org/10.1016/j. drugalcdep.2020.107904.
Taylor, A.N., Tio, D.L., Heng, N.S., and Yirmiya, R., "Alcohol Consumption Attenuates Febrile Responses to Lipopolysaccharide and Interleukin-1 Beta in Male Rats," *Alcoholism Clinical and Experimental Research* 26, no. 1 (2002): 44–52.

254 Zakhari, S., "Overview: How Is Alcohol Metabolized by the Body?" *Alcohol Research & Health* 29, no. 4 (2006): 245–54.
Luo, G., Huang, L., and Zhang, Z., "The Molecular Mechanisms of Acetaminophen-Induced Hepatotoxicity and Its Potential Therapeutic Targets," *Experimental Biology and Medicine (Maywood)* 248, no. 5 (2023): 412–24. https://doi.org/10.1177/15353702221147563.

255 Dungan, A.R., VanRyzin, J.W., Wood, J.A., et al., "Early Life Exposure to Acetaminophen Exerts Cohort-Dependent Effects on Nursing and Play Behavior in Sprague Dawley Rats," *SSRN (Preprint)* http://dx.doi.org/10.2139/ ssrn.5223387 (2025). http://dx.doi.org/10.2139/ssrn.5223387.

256 Daniolou, S., Pandis, N., and Znoj, H., "The Efficacy of Early Interventions for Children with Autism Spectrum Disorders: A Systematic Review and Meta-Analysis," *Journal of Clinical Medicine* 11, no. 17 (2022). https://doi.org/10.3390/jcm11175100.
Reid, N., Dawe, S., Shelton, D., et al., "Systematic Review of Fetal Alcohol Spectrum Disorder Interventions across the Life Span," *Alcoholism Clinical and Experimental Research* 39, no. 12 (2015): 2283–95. https://doi.org/10.1111/acer.12903.

257 Parker, W., Anderson, L.G., Jones, J.P., et al., "The Dangers of Acetaminophen for Neurodevelopment Outweigh Scant Evidence for Long-Term Benefits," *Children* 11, no. 1 (2024): 44. https://www.mdpi.com/2227-9067/11/1/44.

258 Posadas, I., Santos, P., Blanco, A., Muñoz-Fernández, M., and Ceña, V., "Acetaminophen Induces Apoptosis in Rat Cortical Neurons," *PLOS One* 5, no. 12 (2010): e15360. https://doi.org/10.1371/journal.pone.0015360.
Labba, N.-A., Wæhler, H.A., Houdaifi, N., et al., "Paracetamol Perturbs Neuronal Arborization and Disrupts the Cytoskeletal Proteins SPTBN1 and TUBB3 in Both Human and Chicken in Vitro Models," *Toxicology and Applied Pharmacology* 449 (2022): 116130. https://doi.org/https://doi.org/10.1016/j.taap.2022.116130.
Vigo, M.B., Pérez, M.J., De Fino, F., et al., "Acute Acetaminophen Intoxication Induces Direct Neurotoxicity in Rats Manifested as Astrogliosis and Decreased Dopaminergic Markers in Brain Areas Associated with Locomotor Regulation," *Biochemical Pharmacology* 170 (2019): 113662. https://doi.org/https://doi.org/10.1016/j.bcp.2019.113662.
Schultz, S., DeSilva, M., Gu, T.T., Qiang, M., and Whang, K., "Effects of the Analgesic Acetaminophen (Paracetamol) and Its Para-Aminophenol Metabolite on Viability of Mouse-Cultured Cortical Neurons," *Basic & Clinical Pharmacology & Toxicology* 110, no. 2 (2012): 141–44. https://doi.org/10.1111/j.1742-7843.2011.00767.x.

259 Hogestatt, E.D., Jonsson, B.A., Ermund, A., et al., "Conversion of Acetaminophen to the Bioactive N-Acylphenolamine AM_{404} Via Fatty Acid Amide Hydrolase-Dependent Arachidonic Acid Conjugation in the Nervous System," *Journal of Biological Chemistry* 280, no. 36 (2005): 31405–12. https://doi.org/10.1074/jbc.M501489200.
Hinz, B., Cheremina, O., and Brune, K., "Acetaminophen (Paracetamol) Is a Selective Cyclooxygenase-2 Inhibitor in Man," *FASEB Journal* 22, no. 2 (2008): 383–90. https://doi.org/10.1096/fj.07-8506com.

260 Hirai, T., Umeda, N., Harada, T., et al., "Arachidonic Acid-Derived Dihydroxy Fatty Acids in Neonatal Cord Blood Relate Symptoms of Autism Spectrum Disorders and Social Adaptive Functioning: Hamamatsu Birth Cohort for Mothers and Children (HBC Study)," *Psychiatry Clin Neurosci* 78, no. 9 (2024): 546–57. https://doi.org/10.1111/pcn.13710.
El-Ansary, A., Alfawaz, H.A., Bacha, A.B., and Al-Ayadhi, L.Y., "Combining Anti-Mitochondrial Antibodies, Anti-Histone, and PLA2/COX Biomarkers to Increase Their Diagnostic Accuracy for Autism Spectrum Disorders," *Brain Sciences* 14, no. 6 (2024): 576. https://www.mdpi.com/2076-3425/14/6/576.

261 Posadas, I., Santos, P., Blanco, A., Muñoz-Fernández, M., and Ceña, V., "Acetaminophen Induces Apoptosis in Rat Cortical Neurons," *PLOS One* 5, no. 12 (2010): e15360. https://doi.org/10.1371/journal.pone.0015360.

262 Donovan, A.P., and Basson, M.A., "The Neuroanatomy of Autism—A Developmental Perspective," *Journal of Anatomy* 230, no. 1 (2017): 4–15. https://doi.org/10.1111/joa.12542.
Casanova, M.F., Sokhadze, E.M., Casanova, E.L., et al., "Translational Neuroscience in Autism: From Neuropathology to Transcranial Magnetic Stimulation Therapies," *Psychiatric Clinics of North America* 43, no. 2 (2020): 229–48. https://doi.org/10.1016/j.psc.2020.02.004.

263 Salas, V.M., and Corcoran, G.B., "Calcium-Dependent DNA Damage and Adenosine 3',5'-Cyclic Monophosphate-Independent Glycogen Phosphorylase Activation in an in Vitro Model of Acetaminophen-Induced Liver Injury," *Hepatology* 25, no. 6 (1997): 1432–38. https://doi.org/10.1002/hep.510250621.

264 Pourtavakoli, A., and Ghafouri-Fard, S., "Calcium Signaling in Neurodevelopment and Pathophysiology of Autism Spectrum Disorders," *Molecular Biology Reports* 49, no. 11 (2022): 10811–23. https://doi.org/10.1007/s11033-022-07775-6.
Schmunk, G., Nguyen, R.L., Ferguson, D.L., Kumar, K., Parker, I., and Gargus, J.J., "High-Throughput Screen Detects Calcium Signaling Dysfunction in Typical Sporadic Autism Spectrum Disorder," *Scientific Reports* 7 (2017): 40740. https://doi.org/10.1038/srep40740.

265 Parker, W., Anderson, L.G., Jones, J.P., et al., "The Dangers of Acetaminophen for Neurodevelopment Outweigh Scant Evidence for Long-Term Benefits," *Children* 11, no. 1 (2024): 44. https://www.mdpi.com/2227-9067/11/1/44.

266 Jones, J.P., 3rd, Williamson, L., Konsoula, Z., Anderson, R., Reissner, K.J., and Parker, W., "Evaluating the Role of Susceptibility Inducing Cofactors and of Acetaminophen in the Etiology of Autism Spectrum Disorder," *Life (Basel)* 14, no. 8 (2024). https://doi.org/10.3390/life14080918.

267 Jones, J., Konsoula, Z., Williamson, L., Anderson, R., Meza-Keuthen, S., and Parker, W., "Three Mandatory Doses of Acetaminophen During the First Months of Life with the MenB Vaccine: A Protocol for the Induction of Autism Spectrum Disorder in Susceptible Individuals," Preprints, 2025. https://doi.org/10.20944/preprints202501.0319.v5.

268 National Centre for Immunisation Research and Surveillance, "Significant Events in Meningococcal Vaccination Practice in Australia," updated July, 2020, https://www.ncirs.org.au/sites/default/files/2020-07/Meningococcal-history-July%202020.pdf.

269 Immunisation Advisory Centre, "Paracetamol Use with Bexsero," updated 2023, https://www.immune.org.nz/factsheets/paracetamol-use-with-bexsero-r.

270 Immunisation Advisory Centre, "Paracetamol Use with Bexsero," updated 2023, https://www.immune.org.nz/factsheets/paracetamol-use-with-bexsero-r.
Meningococcal Disease (Australian Government, 2024, updated Jan. 23, 2026), https://immunisationhandbook.health.gov.au/contents/vaccine-preventable-diseases/meningococcal-disease.

"Using Paracetamol to Prevent and Treat Fever after MenB Vaccination," National Health Service, 2026, https://assets.publishing.service.gov.uk/media/634469858fa8f52a5b78cdf8/UKHSA-paracetamol-MenB-2022.pdf.

"Meningococcal Vaccines: Canadian Immunization Guide," Government of Canada, updated Aug. 14, 2025, https://www.canada.ca/en/public-health/services/publications/healthy-living/canadian-immunization-guide-part-4-active-vaccines/page-13-meningococcal-vaccine.html.

Ministry of Health (Israel), "Vaccines for Children," Nov. 15, 2021, https://www.gov.il/en/pages/vaccines-for-children-pamphlet.

271 Immunisation Advisory Centre, "Paracetamol Use with Bexsero," updated 2023, https://www.immune.org.nz/factsheets/paracetamol-use-with-bexsero-r.

272 Parker, W., Anderson, L.G., Jones, J.P., et al., "The Dangers of Acetaminophen for Neurodevelopment Outweigh Scant Evidence for Long-Term Benefits," *Children* 11, no. 1 (2024): 44. https://www.mdpi.com/2227-9067/11/1/44.

Jones, J.P., 3rd, Williamson, L., Konsoula, Z., Anderson, R., Reissner, K.J., and Parker, W., "Evaluating the Role of Susceptibility Inducing Cofactors and of Acetaminophen in the Etiology of Autism Spectrum Disorder," *Life (Basel)* 14, no. 8 (2024). https://doi.org/10.3390/life14080918.

273 Bilbo, S.D., Nevison, C.D., and Parker, W., "A Model for the Induction of Autism in the Ecosystem of the Human Body: The Anatomy of a Modern Pandemic?" *Microbial Ecology in Health and Disease* 26 (2015): 26253. https://doi.org/10.3402/mehd.v26.26253.

Chapter 8: Little Scientific Blunders Support the Big Scientific Blunders

274 Sheikh, J., Allotey, J., Sobhy, S., et al., "Maternal Paracetamol (Acetaminophen) Use During Pregnancy and Risk of Autism Spectrum Disorder and Attention Deficit/Hyperactivity Disorder in Offspring: Umbrella Review of Systematic Reviews," *BMJ* 391 (2025): e088141. https://doi.org/10.1136/bmj-2025-088141.

Editor, "Weaponizing Uncertainty in Science and in Public Health Puts People in Harm's Way," *Nature* 646, no. 8084 (2025): 259–60. https://doi.org/10.1038/d41586-025-03167-5.

Pearson, H., and Ledford, H., "Trump Links Autism and Tylenol: Is There Any Truth to It?" *Nature* 646, no. 8083 (2025): 13–14. https://doi.org/10.1038/d41586-025-02876-1.

Looi, M.K., and Bowie, K., "Autism: Trump Links Condition to Tylenol and Touts Leucovorin as 'First' US Therapeutic," *BMJ* 390 (2025): r2004. https://doi.org/10.1136/bmj.r2004.

Schweitzer, K., "Acetaminophen Use in Pregnancy-Study Author Explains the Data," *JAMA* 334, no. 17 (2025): 1499–501. https://doi.org/10.1001/jama.2025.19345.

Andrade, C., "Maternal Use of Acetaminophen (Paracetamol) During Pregnancy and Neurodevelopmental Disorders in Offspring: A Reasoned Evaluation of Risk," *Journal of Clinical Psychiatry* 86, no. 4 (2025). https://doi.org/10.4088/JCP.25f16187.

Louwen, F., Deuster, E., McAuliffe, F.M., et al., "Paracetamol (Acetaminophen) Use During Pregnancy and Autism Risk: Evidence Does Not Support Causal Association," *International Journal of Gynaecology & Obstetrics* 171, no. 3 (2025): 915–19. https://doi.org/10.1002/ijgo.70577.

Wise, J., "Paracetamol (Tylenol): No Clear Link between Use in Pregnancy and Autism or ADHD in Children, Rapid Review Finds," *BMJ* 391 (2025): r2368. https://doi.org/10.1136/bmj.r2368.

Lee, B.K., Stephansson, O., and Gardner, R.M., "Paracetamol (Acetaminophen) Use in Pregnancy and Risk of Autism and ADHD," *BMJ* 391 (2025): r2438. https://doi.org/10.1136/bmj.r2438.

Lang, K., "Trump's Claims on Tylenol (Paracetamol), Vaccines, and Autism—What's the Truth?" *BMJ* 390 (2025): r2025. https://doi.org/10.1136/bmj.r2025.

Aronson, J.K., "When I Use a Word . . . Paracetamol/Acetaminophen-Autism and Asthma," *BMJ* 390 (2025): r2032. https://doi.org/10.1136/bmj.r2032.

Gupta, S., Kopacz, K.S., and Hurdle, M.F.B., "Acetaminophen, Autism, and Analgesia in Pregnancy—A Pain Medicine Perspective," *Pain Management* 16, no. 1 (2026): 1–3. https://doi.org/10.1080/17581869.2025.2588240.

Ottewell, L., and Wright, F., "Purity, Politics and Pain: Trump's Paracetamol Posturing and the Moralisation of Pregnancy," *Medical Humanities* (2025). https://doi.org/10.1136/medhum-2025-013652.

275 Sheikh, J., Allotey, J., Sobhy, S., et al., "Maternal Paracetamol (Acetaminophen) Use During Pregnancy and Risk of Autism Spectrum Disorder and Attention Deficit/Hyperactivity Disorder in Offspring: Umbrella Review of Systematic Reviews," *BMJ* 391 (2025): e088141. https://doi.org/10.1136/bmj-2025-088141.

Editor, "Weaponizing Uncertainty in Science and in Public Health Puts People in Harm's Way," *Nature* 646, no. 8084 (2025): 259–60. https://doi.org/10.1038/d41586-025-03167-5.

Pearson, H., and Ledford, H., "Trump Links Autism and Tylenol: Is There Any Truth to It?" *Nature* 646, no. 8083 (2025): 13–14. https://doi.org/10.1038/d41586-025-02876-1.

Looi, M.K., and Bowie, K., "Autism: Trump Links Condition to Tylenol and Touts Leucovorin as 'First' US Therapeutic," *BMJ* 390 (2025): r2004. https://doi.org/10.1136/bmj.r2004.

Schweitzer, K., "Acetaminophen Use in Pregnancy-Study Author Explains the Data," *JAMA* 334, no. 17 (2025): 1499–501. https://doi.org/10.1001/jama.2025.19345.

Andrade, C., "Maternal Use of Acetaminophen (Paracetamol) During Pregnancy and Neurodevelopmental Disorders in Offspring: A Reasoned Evaluation of Risk," *Journal of Clinical Psychiatry* 86, no. 4 (2025). https://doi.org/10.4088/JCP.25f16187.

Louwen, F., Deuster, E., McAuliffe, F.M., et al., "Paracetamol (Acetaminophen) Use During Pregnancy and Autism Risk: Evidence Does Not Support Causal Association," *International Journal of Gynaecology & Obstetrics* 171, no. 3 (2025): 915–19. https://doi.org/10.1002/ijgo.70577.

Wise, J., "Paracetamol (Tylenol): No Clear Link between Use in Pregnancy and Autism or ADHD in Children, Rapid Review Finds," *BMJ* 391 (2025): r2368. https://doi.org/10.1136/bmj.r2368.
Lee, B.K., Stephansson, O., and Gardner, R.M., "Paracetamol (Acetaminophen) Use in Pregnancy and Risk of Autism and ADHD," *BMJ* 391 (2025): r2438. https://doi.org/10.1136/bmj.r2438.
Lang, K., "Trump's Claims on Tylenol (Paracetamol), Vaccines, and Autism—What's the Truth?" *BMJ* 390 (2025): r2025. https://doi.org/10.1136/bmj.r2025.
Aronson, J.K., "When I Use a Word . . . Paracetamol/Acetaminophen-Autism and Asthma," *BMJ* 390 (2025): r2032. https://doi.org/10.1136/bmj.r2032.
Gupta, S., Kopacz, K.S., and Hurdle, M.F.B., "Acetaminophen, Autism, and Analgesia in Pregnancy—A Pain Medicine Perspective," *Pain Management* 16, no. 1 (2026): 1–3. https://doi.org/10.1080/17581869.2025.2588240.
Ottewell, L., and Wright, F., "Purity, Politics and Pain: Trump's Paracetamol Posturing and the Moralisation of Pregnancy," *Medical Humanities* (2025). https://doi.org/10.1136/medhum-2025-013652.

276 Maher, B., "Personal Genomes: The Case of the Missing Heritability," *Nature* 456, no. 7218 (2008): 18–21. https://doi.org/10.1038/456018a.

277 Zhao, L., Jones, J., Anderson, L., et al., "Acetaminophen Causes Neurodevelopmental Injury in Susceptible Babies and Children: No Valid Rationale for Controversy," *Clinical and Experimental Pediatrics* (2023). https://doi.org/10.3345/cep.2022.01319.

278 Schultz, S.T., Klonoff-Cohen, H.S., Wingard, D.L., Akshoomoff, N.A., Macera, C.A., and Ji, M., "Acetaminophen (Paracetamol) Use, Measles-Mumps-Rubella Vaccination, and Autistic Disorder. The Results of a Parent Survey," *Autism* 12, no. 3 (2008): 293–307.

279 Good, P., "Did Acetaminophen Provoke the Autism Epidemic?" *Alternative Medicine Review* 14, no. 4 (2009): 364–72.

280 Good, P., "Did Acetaminophen Provoke the Autism Epidemic?" *Alternative Medicine Review* 14, no. 4 (2009): 364–72.

281 Parker, W., Corrigan, P.T., Anderson, R., et al., "Evidence That Acetaminophen Triggers Autism in Susceptible Individuals Has Been Ignored and Mishandled for More Than a Decade," *Journal of the Academy of Public Health* (2025). https://publichealth.realclearjournals.org/literature-syntheses/2025/10/evidence-that-acetaminophen-triggers-autism-in-susceptible-individuals-has-been-ignored-and-mishandled-for-more-than-a-decade/.

282 Frisch, M., and Simonsen, J., "Ritual Circumcision and Risk of Autism Spectrum Disorder in 0- to 9-Year-Old Boys: National Cohort Study in Denmark," *Journal of the Royal Society of Medicine* 108, no. 7 (2015): 266–79. https://doi.org/10.1177/0141076814565942.

283 Miani, A., Di Bernardo, G.A., Højgaard, A.D., et al., "Neonatal Male Circumcision Is Associated with Altered Adult Socio-Affective Processing," *Heliyon* 6, no. 11 (2020): e05566. https://doi.org/10.1016/j.heliyon.2020.e05566

284 Ji, Y., Azuine, R.E., Zhang, Y., et al., "Association of Cord Plasma Biomarkers of in Utero Acetaminophen Exposure with Risk of Attention-Deficit/Hyperactivity

Disorder and Autism Spectrum Disorder in Childhood," *JAMA Psychiatry* 77, no. 2 (2020): 180–89. https://doi.org/10.1001/jamapsychiatry.2019.3259.

285 Ali, N.A., Kennon-McGill, S., Parker, L.D., James, L.P., Fantegrossi, W.E., and McGill, M.R., "NAPQI Is Absent in the Mouse Brain after Sub-Hepatotoxic and Hepatotoxic Doses of Acetaminophen," *Toxicological Sciences* 205, no. 2 (2025): 274–78. https://doi.org/10.1093/toxsci/kfaf034.

286 Ali, N.A., Kennon-McGill, S., Parker, L.D., James, L.P., Fantegrossi, W.E., and McGill, M.R., "NAPQI Is Absent in the Mouse Brain after Sub-Hepatotoxic and Hepatotoxic Doses of Acetaminophen," *Toxicological Sciences* 205, no. 2 (2025): 274–78. https://doi.org/10.1093/toxsci/kfaf034.

287 Viberg, H., Eriksson, P., Gordh, T., and Fredriksson, A., "Paracetamol (Acetaminophen) Administration During Neonatal Brain Development Affects Cognitive Function and Alters Its Analgesic and Anxiolytic Response in Adult Male Mice," *Toxicological Sciences* 138, no. 1 (2013): 139–47. https://doi .org/10.1093/toxsci/kft329.
Philippot, G., Gordh, T., Fredriksson, A., and Viberg, H., "Adult Neurobehavioral Alterations in Male and Female Mice Following Developmental Exposure to Paracetamol (Acetaminophen): Characterization of a Critical Period," *Journal of Applied Toxicology* 37, no. 10 (2017): 1174–81. https://doi .org/10.1002/jat.3473.

288 Viberg, H., Eriksson, P., Gordh, T., and Fredriksson, A., "Paracetamol (Acetaminophen) Administration During Neonatal Brain Development Affects Cognitive Function and Alters Its Analgesic and Anxiolytic Response in Adult Male Mice," *Toxicological Sciences* 138, no. 1 (2013): 139–47. https://doi .org/10.1093/toxsci/kft329.
Philippot, G., Gordh, T., Fredriksson, A., and Viberg, H., "Adult Neurobehavioral Alterations in Male and Female Mice Following Developmental Exposure to Paracetamol (Acetaminophen): Characterization of a Critical Period," *Journal of Applied Toxicology* 37, no. 10 (2017): 1174–81. https://doi .org/10.1002/jat.3473.

289 Posadas, I., Santos, P., Blanco, A., Muñoz-Fernández, M., and Ceña, V., "Acetaminophen Induces Apoptosis in Rat Cortical Neurons," *PLOS One* 5, no. 12 (2010): e15360. https://doi.org/10.1371/journal.pone.0015360.

290 Heard, K.J., Green, J.L., James, L.P., et al., "Acetaminophen-Cysteine Adducts During Therapeutic Dosing and Following Overdose," *BMC Gastroenterology* 11 (2011): 20. https://doi.org/10.1186/1471-230x-11-20.

291 Ali, N.A., Kennon-McGill, S., Parker, L.D., James, L.P., Fantegrossi, W.E., and McGill, M.R., "NAPQI Is Absent in the Mouse Brain after Sub-Hepatotoxic and Hepatotoxic Doses of Acetaminophen," *Toxicological Sciences* 205, no. 2 (2025): 274–78. https://doi.org/10.1093/toxsci/kfaf034.

292 Heard, K.J., Green, J.L., James, L.P., et al., "Acetaminophen-Cysteine Adducts During Therapeutic Dosing and Following Overdose," *BMC Gastroenterology* 11 (2011): 20. https://doi.org/10.1186/1471-230x-11-20.

293 Dungan, A.R., VanRyzin, J.W., Wood, J.A., et al., "Early Life Exposure to Acetaminophen Exerts Cohort-Dependent Effects on Nursing and Play

Behavior in Sprague Dawley Rats," *SSRN (Preprint)* http://dx.doi.org/10.2139/ssrn.5223387 (2025). http://dx.doi.org/10.2139/ssrn.5223387.

294 Andrade, C., "Harking, Cherry-Picking, P-Hacking, Fishing Expeditions, and Data Dredging and Mining as Questionable Research Practices," *Journal of Clinical Psychiatry* 82, no. 1 (2021). https://doi.org/10.4088/JCP.20f13804.

295 Kennon-McGill, S., and McGill, M.R., "Extrahepatic Toxicity of Acetaminophen: Critical Evaluation of the Evidence and Proposed Mechanisms," *Journal of Clinical and Translational Research* 3, no. 3 (2018): 297–310. https://doi.org/10.18053/jctres.03.201703.005.

296 Constantino, J.N., and Marrus, N., "The Early Origins of Autism," *Child and Adolescent Psychiatric Clinics of North America* 26, no. 3 (2017): 555–70. https://doi.org/10.1016/j.chc.2017.02.008.

297 Ali, N.A., Kennon-McGill, S., Parker, L.D., James, L.P., Fantegrossi, W.E., and McGill, M.R., "NAPQI Is Absent in the Mouse Brain after Sub-Hepatotoxic and Hepatotoxic Doses of Acetaminophen," *Toxicological Sciences* 205, no. 2 (2025): 274–78. https://doi.org/10.1093/toxsci/kfaf034.

298 Beinart, D., Ren, D., Pi, C., et al., "Immunization Enhances the Natural Antibody Repertoire," *EXCLI Journal* 16 (2017): 1018–30. https://doi.org/doi.org/10.17179/excli2017-500.

299 Yu, P.B., Parker, W., Everett, M.L., Fox, I.J., and Platt, J.L., "Immunochemical Properties of Anti-Gal Alpha 1-3gal Antibodies after Sensitization with Xenogeneic Tissues," *Journal of Clinical Immunology* 19, no. 2 (1999): 116–26.
Yu, P.B., Parker, W., Nayak, J.V., and Platt, J.L., "Sensitization with Xenogeneic Tissues Alters the Heavy Chain Repertoire of Human Anti-Galalpha1-3gal Antibodies," *Transplantation* 80, no. 1 (2005): 102–9.

300 Schultz, S., *Understanding Autism: My Quest for Nathan* (Schultz Publishing LLC, 2013).

301 Laurin, M., Everett, M.L., and Parker, W., "The Cecal Appendix: One More Immune Component with a Function Disturbed by Post-Industrial Culture," *Anatomical Record* 294, no. 4 (2011): 567–79. https://doi.org/10.1002/ar.21357.
Devalapalli, A.P., Lesher, A., Shieh, K., et al., "Increased Levels of IgE and Autoreactive, Polyreactive IgG in Wild Rodents: Implications for the Hygiene Hypothesis," *Scandinavian Journal of Immunology* 64 (2006): 125–36. https://doi.org/10.1111/j.1365-3083.2006.01785.x.
Lesher, A., Li, B., Whitt, P., et al., "Increased Il-4 Production and Attenuated Proliferative and Proinflammatory Responses of Splenocytes from Wild-Caught Rats (Rattus Norvegicus)," *Immunology and Cell Biology* 84 (2006): 374–82.
Lin, S.S., Holzknecht, Z.E., Trama, A.M., et al., "Immune Characterization of Wild-Caught *Rattus Norvegicus* Suggests Diversity of Immune Activity in Biome-Normal Environments," *Journal of Evolutionary Medicine* 1 (2012): 1–16.
Trama, A.M., Holzknecht, Z.E., Thomas, A.D., et al., "Lymphocyte Phenotypes in Wild-Caught Rats Suggest Potential Mechanisms Underlying Increased Immune Sensitivity in Post-Industrial Environments," *Cellular & Molecular Immunology* 9 (2012): 163–74.

Cheng, A.M., Jaint, D., Thomas, S., Wilson, J., and Parker, W., "Overcoming Evolutionary Mismatch by Self-Treatment with Helminths: Current Practices and Experience," *Journal of Evolutionary Medicine* 3 (2015): Article ID 235910.

Liu, J., Morey, R.A., Wilson, J.K., and Parker, W., "Practices and Outcomes of Self-Treatment with Helminths Based on Physicians' Observations," *Journal of Helminthology* FirstView (2016): 1–11.

Williamson, L.L., McKenney, E.A., Holzknecht, Z.E., et al., "Got Worms? Perinatal Exposure to Helminths Prevents Persistent Immune Sensitization and Cognitive Dysfunction Induced by Early-Life Infection," *Brain, Behavior, and Immunity* 51 (2016): 14–28. https://doi.org/10.1016/j.bbi.2015.07.006.

Parker, W., "Not Infection with Parasitic Worms, but Rather Colonization with Therapeutic Helminths," *Immunology Letters* (2017). https://doi.org/http://dx.doi.org/10.1016/j.imlet.2017.07.008.

Sobotková, K., Parker, W., Levá, J., Růžková, J., Lukeš, J., and Jirků Pomajbíková, K., "Helminth Therapy—from the Parasite Perspective," *Trends in Parasitology* 35, no. 7 (2019): 501–15. https://doi.org/10.1016/j.pt.2019.04.009.

Parker, W., Sarafian, J.T., Broverman, S.A., and Laman, J.D., "Between a Hygiene Rock and a Hygienic Hard Place: Avoiding SARS-CoV-2 While Needing Environmental Exposures for Immunity," *Evolution, Medicine and Public Health* 9, no. 1 (2021): 120–30. https://doi.org/10.1093/emph/eoab006.

Engelenburg, H.J., Lucassen, P.J., Sarafian, J.T., Parker, W., and Laman, J.D., "Multiple Sclerosis and the Microbiota: Progress in Understanding the Contribution of the Gut Microbiome to Disease," *Evolution, Medicine, and Public Health* 10, no. 1 (2022): 277–94. https://doi.org/10.1093/emph/eoac009.

Parker, W., Patel, E., Jirků-Pomajbíková, K., and Laman, J.D., "COVID-19 Morbidity in Lower Versus Higher Income Populations Underscores the Need to Restore Lost Biodiversity of Eukaryotic Symbionts," *iScience* 26, no. 3 (2023): 106167. https://doi.org/10.1016/j.isci.2023.106167.

Parker, W., Jirků, K., Patel, E., Williamson, L., Anderson, L., and Laman, J.D., "Reevaluating Biota Alteration: Reframing Environmental Influences on Chronic Immune Disorders and Exploring Novel Therapeutic Opportunities," *Yale Journal of Biology and Medicine* 97, no. 2 (2024): 253–63. https://doi.org/10.59249/vunf1315.

Schoenecker, J.G., Johnson, R.K., Lesher, A.P., et al., "Exposure of Mice to Topical Bovine Thrombin Induces Systemic Autoimmunity," *American Journal of Pathology* 159, no. 5 (2001): 1957–69.

Bilbo, S.D., Wray, G.A., Perkins, S.E., and Parker, W., "Reconstitution of the Human Biome as the Most Reasonable Solution for Epidemics of Allergic and Autoimmune Diseases," *Medical Hypotheses* 77, no. 4 (2011): 494–504. https://doi.org/doi:10.1016/j.mehy.2011.06.019.

Parker, W., Perkins, S.E., Harker, M., and Muehlenbein, M.P., "A Prescription for Clinical Immunology: The Pills Are Available and Ready for Testing," *Current Medical Research and Opinion* 28 (2012): 1193–202. https://doi.org/10.1185/03007995.2012.695731.

Parker, W., and Ollerton, J., "Evolutionary Biology and Anthropology Suggest Biome Reconstitution as a Necessary Approach toward Dealing with Immune

Disorders," *Evolution, Medicine, and Public Health* 2013 (2013): 89–103. https://doi.org/10.1093/emph/eot008.

Parker, W., "The 'Hygiene Hypothesis' for Allergic Disease Is a Misnomer," *BMJ* 349 (2014): g5267. https://doi.org/10.1136/bmj.g5267.

Brenner, S.L., Jones, J.P., Rutanen-Whaley, R.H., Parker, W., Flinn, M.V., and Muehlenbein, M.P., "Evolutionary Mismatch and Chronic Psychological Stress," *Journal of Evolutionary Medicine* 3 (2015): art235885. https://doi.org/10.4303/jem/235885.

Bono-Lunn, D., Villeneuve, C., Abdulhay, N.J., Harker, M., and Parker, W., "Policy and Regulations in Light of the Human Body as a 'Superorganism' Containing Multiple, Intertwined Symbiotic Relationships," *Clinical Research and Regulatory Affairs* 33, no. 2–4 (2016): 39–48. https://doi.org/10.1080/10601333.2016.1210159.

Kou, H.H., and Parker, W., "Intestinal Worms Eating Neuropsychiatric Disorders? Apparently So," *Brain Research* 1693, no. Pt B (2018): 218–21. https://doi.org/10.1016/j.brainres.2018.01.023.

Villeneuve, C., Kou, H.H., Eckermann, H., et al., "Evolution of the Hygiene Hypothesis into Biota Alteration Theory: What Are the Paradigms and Where Are the Clinical Applications?" *Microbes and Infection* 20, no. 3 (2018): 147–55. https://doi.org/10.1016/j.micinf.2017.11.001.

302 Parker, W., Hornik, C.D., Bilbo, S., et al., "The Role of Oxidative Stress, Inflammation and Acetaminophen Exposure from Birth to Early Childhood in the Induction of Autism," *Journal of International Medical Research* 45, no. 2 (2017): 407–38.

303 Rimland, B., "The Autism Increase: Research Needed on the Vaccine Connection," *Autism Research Review International* 14, no. 1 (2000): 3.

304 Rimland, B., "The Autism Increase: Research Needed on the Vaccine Connection," *Autism Research Review International* 14, no. 1 (2000): 3.

305 Schultz, S.T., Klonoff-Cohen, H.S., Wingard, D.L., Akshoomoff, N.A., Macera, C.A., and Ji, M., "Acetaminophen (Paracetamol) Use, Measles-Mumps-Rubella Vaccination, and Autistic Disorder. The Results of a Parent Survey," *Autism* 12, no. 3 (2008): 293–307.

306 Kern, J.K., Geier, D.A., Sykes, L.K., Haley, B.E., and Geier, M.R., "The Relationship between Mercury and Autism: A Comprehensive Review and Discussion," *Journal of Trace Elements in Medicine and Biology* 37 (2016): 8–24. https://doi.org/10.1016/j.jtemb.2016.06.002.

307 Abu Kuwaik, G., Roberts, W., Zwaigenbaum, L., et al., "Immunization Uptake in Younger Siblings of Children with Autism Spectrum Disorder," *Autism* 18, no. 2 (2014): 148–55. https://doi.org/10.1177/1362361312459111.

308 McFadden, J., "Razor Sharp: The Role of Occam's Razor in Science," *Annals of the New York Academy of Sciences* 1530, no. 1 (2023): 8–17. https://doi.org/https://doi.org/10.1111/nyas.15086.

309 Parker, W., Hornik, C.D., Bilbo, S., et al., "The Role of Oxidative Stress, Inflammation and Acetaminophen Exposure from Birth to Early Childhood in the Induction of Autism," *Journal of International Medical Research* 45, no. 2 (2017): 407–38.

310 Good, P., "Evidence the U.S. Autism Epidemic Initiated by Acetaminophen (Tylenol) Is Aggravated by Oral Antibiotic Amoxicillin/Clavulanate (Augmentin) and Now Exponentially by Herbicide Glyphosate (Roundup)," *Clinical Nutrition ESPEN* 23 (2018): 171–83. https://doi.org/10.1016/j.clnesp.2017.10.005.

311 Shaw, W., "Hypothesis: 2 Major Environmental and Pharmaceutical Factors—Acetaminophen Exposure and Gastrointestinal Overgrowth of Clostridia Bacteria Induced by Ingestion of Glyphosate-Contaminated Foods-Dysregulate the Developmental Protein Sonic Hedgehog and Are Major Causes of Autism," *Integrative Medicine (Encinitas)* 23, no. 3 (2024): 12–23.

312 Parker, W., Hornik, C.D., Bilbo, S., et al., "The Role of Oxidative Stress, Inflammation and Acetaminophen Exposure from Birth to Early Childhood in the Induction of Autism," *Journal of International Medical Research* 45, no. 2 (2017): 407–38.

Chapter 9: Emotion-Based Objections to the Scientific Evidence

313 Cendejas-Hernandez, J., Sarafian, J., Lawton, V., et al., "Paracetamol (Acetaminophen) Use in Infants and Children Was Never Shown to Be Safe for Neurodevelopment: A Systematic Review with Citation Tracking," *European Journal of Pediatrics* 181 (2022): 1835–57. https://doi.org/10.1007/s00431-022-04407-w.

314 Sultan, M., Tump, A.N., Ehmann, N., et al., "Susceptibility to Online Misinformation: A Systematic Meta-Analysis of Demographic and Psychological Factors," *Proceedings of the National Academy of Sciences* 121, no. 47 (2024): e2409329121. https://doi.org/doi:10.1073/pnas.2409329121.
Kahan, D., Peters, E., Dawson, E., and Slovic, P., "Motivated Numeracy and Enlightened Self-Government," *Behavioural Public Policy* 1 (2017): 54–86. https://doi.org/10.1017/bpp.2016.2.

315 Mazzei, M.J., DeBode, J., Gangloff, K.A., and Song, R., "Old Habits Die Hard: A Review and Assessment of the Threat-Rigidity Literature," *Journal of Management* 51, no. 6 (2025): 2154–81. https://doi.org/10.1177/01492063241286493.

316 Asch, S.E., "Studies of Independence and Conformity: I. A Minority of One against a Unanimous Majority," *Psychological Monographs: General and Applied* 70, no. 9 (1956): 1–70. https://doi.org/10.1037/h0093718.
Capuano, C., and Chekroun, P., "A Systematic Review of Research on Conformity," *International Review of Social Psychology* 37 (2024): 13. https://doi.org/10.5334/irsp.874.

317 Mazzei, M.J., DeBode, J., Gangloff, K.A., and Song, R., "Old Habits Die Hard: A Review and Assessment of the Threat-Rigidity Literature," *Journal of Management* 51, no. 6 (2025): 2154–81. https://doi.org/10.1177/01492063241286493.

318 Sultan, M., Tump, A.N., Ehmann, N., et al., "Susceptibility to Online Misinformation: A Systematic Meta-Analysis of Demographic and Psychological Factors," *Proceedings of the National Academy of Sciences* 121, no. 47 (2024): e2409329121. https://doi.org/doi:10.1073/pnas.2409329121.

Kahan, D., Peters, E., Dawson, E., and Slovic, P., "Motivated Numeracy and Enlightened Self-Government," *Behavioural Public Policy* 1 (2017): 54–86. https://doi.org/10.1017/bpp.2016.2.

319 Jones, J., Konsoula, Z., Williamson, L., Anderson, R., Meza-Keuthen, S., and Parker, W., "Three Mandatory Doses of Acetaminophen During the First Months of Life with the Menb Vaccine: A Protocol for the Induction of Autism Spectrum Disorder in Susceptible Individuals," Preprints, 2025. https://doi.org/10.20944/preprints202501.0319.v5.

320 Jones, J., Konsoula, Z., Williamson, L., Anderson, R., Meza-Keuthen, S., and Parker, W., "Three Mandatory Doses of Acetaminophen During the First Months of Life with the MenB Vaccine: A Protocol for the Induction of Autism Spectrum Disorder in Susceptible Individuals," Preprints, 2025. https://doi.org/10.20944/preprints202501.0319.v5.
Kahan, D., Peters, E., Dawson, E., and Slovic, P., "Motivated Numeracy and Enlightened Self-Government," *Behavioural Public Policy* 1 (2017): 54–86. https://doi.org/10.1017/bpp.2016.2.

321 Asch, S.E., "Studies of Independence and Conformity: I. A Minority of One against a Unanimous Majority," *Psychological Monographs: General and Applied* 70, no. 9 (1956): 1–70. https://doi.org/10.1037/h0093718.
Capuano, C., and Chekroun, P., "A Systematic Review of Research on Conformity," *International Review of Social Psychology* 37 (2024): 13. https://doi.org/10.5334/irsp.874.

322 Grzyb, T., Dolinski, D., Sudoł-Malisz, M., Kulik, G., and Mielczarek, Ł., "Doubting the Power of Prestige: Obedience to Authority Beyond Institutional and Research Justifications," *Scientific Reports* 15, no. 1 (2025): 24911. https://doi.org/10.1038/s41598-025-10331-4.

323 McFadden, J., "Razor Sharp: The Role of Occam's Razor in Science," *Annals of the New York Academy of Sciences* 1530, no. 1 (2023): 8–17. https://doi.org/https://doi.org/10.1111/nyas.15086.

Chapter 10: Conflicts of Interest in the Medical and Scientific Community—Too Much Was Bet Too Soon

324 Grinker, R.R., "Offit Paul: Autism's False Prophets: Bad Science, Risky Medicine, and the Search for a Cure," *Journal of Autism and Developmental Disorders* 39, no. 3 (2009): 544–46. https://doi.org/10.1007/s10803-008-0679-y.

325 Schultz, S.T., Klonoff-Cohen, H.S., Wingard, D.L., Akshoomoff, N.A., Macera, C.A., and Ji, M., "Acetaminophen (Paracetamol) Use, Measles-Mumps-Rubella Vaccination, and Autistic Disorder. The Results of a Parent Survey," *Autism* 12, no. 3 (2008): 293–307.

326 Parker, W., Corrigan, P.T., Anderson, R., et al., "Evidence That Acetaminophen Triggers Autism in Susceptible Individuals Has Been Ignored and Mishandled for More Than a Decade," *Journal of the Academy of Public Health* (2025). https://publichealth.realclearjournals.org/literature-syntheses/2025/10/evidence-that-acetaminophen-triggers-autism-in-susceptible-individuals-has-been-ignored-and-mishandled-for-more-than-a-decade/.

Good, P., "Did Acetaminophen Provoke the Autism Epidemic?" *Alternative Medicine Review* 14, no. 4 (2009): 364–72.

327 Viberg, H., Eriksson, P., Gordh, T., and Fredriksson, A., "Paracetamol (Acetaminophen) Administration During Neonatal Brain Development Affects Cognitive Function and Alters Its Analgesic and Anxiolytic Response in Adult Male Mice," *Toxicological Sciences* 138, no. 1 (2013): 139–47. https://doi.org/10.1093/toxsci/kft329.

328 Parker, W., Hornik, C.D., Bilbo, S., et al., "The Role of Oxidative Stress, Inflammation and Acetaminophen Exposure from Birth to Early Childhood in the Induction of Autism," *Journal of International Medical Research* 45, no. 2 (2017): 407–38.

329 Ahlqvist, V.H., Sjöqvist, H., Dalman, C., et al., "Acetaminophen Use During Pregnancy and Children's Risk of Autism, ADHD, and Intellectual Disability," *JAMA* 331, no. 14 (2024): 1205–14. https://doi.org/10.1001/jama.2024.3172.

330 Editor, "Weaponizing Uncertainty in Science and in Public Health Puts People in Harm's Way," *Nature* 646, no. 8084 (2025): 259–60. https://doi.org/10.1038/d41586-025-03167-5.
Pearson, H., and Ledford, H., "Trump Links Autism and Tylenol: Is There Any Truth to It?" *Nature* 646, no. 8083 (2025): 13–14. https://doi.org/10.1038/d41586-025-02876-1.
Looi, M.K., and Bowie, K., "Autism: Trump Links Condition to Tylenol and Touts Leucovorin as 'First' US Therapeutic," *BMJ* 390 (2025): r2004. https://doi.org/10.1136/bmj.r2004.
Schweitzer, K., "Acetaminophen Use in Pregnancy-Study Author Explains the Data," *JAMA* 334, no. 17 (2025): 1499–501. https://doi.org/10.1001/jama.2025.19345.
Andrade, C., "Maternal Use of Acetaminophen (Paracetamol) During Pregnancy and Neurodevelopmental Disorders in Offspring: A Reasoned Evaluation of Risk," *Journal of Clinical Psychiatry* 86, no. 4 (2025). https://doi.org/10.4088/JCP.25f16187.
Louwen, F., Deuster, E., McAuliffe, F.M., et al., "Paracetamol (Acetaminophen) Use During Pregnancy and Autism Risk: Evidence Does Not Support Causal Association," *International Journal of Gynaecology & Obstetrics* 171, no. 3 (2025): 915–19. https://doi.org/10.1002/ijgo.70577.

331 Parker, W., Anderson, L.G., Jones, J.P., et al., "The Dangers of Acetaminophen for Neurodevelopment Outweigh Scant Evidence for Long-Term Benefits," *Children* 11, no. 1 (2024): 44. https://www.mdpi.com/2227-9067/11/1/44.

332 Ahlqvist, V.H., Sjöqvist, H., Dalman, C., et al., "Acetaminophen Use During Pregnancy and Children's Risk of Autism, ADHD, and Intellectual Disability," *JAMA* 331, no. 14 (2024): 1205–14. https://doi.org/10.1001/jama.2024.3172.

333 Parker, W., Hornik, C.D., Bilbo, S., et al., "The Role of Oxidative Stress, Inflammation and Acetaminophen Exposure from Birth to Early Childhood in the Induction of Autism," *Journal of International Medical Research* 45, no. 2 (2017): 407–38.

334 Jones, J.P., 3rd, Williamson, L., Konsoula, Z., Anderson, R., Reissner, K.J., and Parker, W., "Evaluating the Role of Susceptibility Inducing Cofactors and of

Acetaminophen in the Etiology of Autism Spectrum Disorder," *Life (Basel)* 14, no. 8 (2024). https://doi.org/10.3390/life14080918.

335 Patel, E., Jones, J.P., 3rd, Bono-Lunn, D., et al., "The Safety of Pediatric Use of Paracetamol (Acetaminophen): A Narrative Review of Direct and Indirect Evidence," *Minerva Pediatrics (Torino)* 74, no. 6 (2022): 774–88. https://doi .org/10.23736/s2724-5276.22.06932-4.

336 Jones, J.P., 3rd, Williamson, L., Konsoula, Z., Anderson, R., Reissner, K.J., and Parker, W., "Evaluating the Role of Susceptibility Inducing Cofactors and of Acetaminophen in the Etiology of Autism Spectrum Disorder," *Life (Basel)* 14, no. 8 (2024). https://doi.org/10.3390/life14080918.

337 Jones, J.P., 3rd, Williamson, L., Konsoula, Z., Anderson, R., Reissner, K.J., and Parker, W., "Evaluating the Role of Susceptibility Inducing Cofactors and of Acetaminophen in the Etiology of Autism Spectrum Disorder," *Life (Basel)* 14, no. 8 (2024). https://doi.org/10.3390/life14080918.

338 Parker, W., Anderson, L.G., Jones, J.P., et al., "The Dangers of Acetaminophen for Neurodevelopment Outweigh Scant Evidence for Long-Term Benefits," *Children* 11, no. 1 (2024): 44. https://www.mdpi.com/2227-9067/11/1/44.

339 Ji, Y., Azuine, R.E., Zhang, Y., et al., "Association of Cord Plasma Biomarkers of in Utero Acetaminophen Exposure with Risk of Attention-Deficit/Hyperactivity Disorder and Autism Spectrum Disorder in Childhood," *JAMA Psychiatry* 77, no. 2 (2020): 180–89. https://doi.org/10.1001/jamapsychiatry.2019.3259.

340 Zhao, L., Jones, J., Anderson, L., et al., "Acetaminophen Causes Neurodevelopmental Injury in Susceptible Babies and Children: No Valid Rationale for Controversy," *Clinical and Experimental Pediatrics* (2023). https:// doi.org/10.3345/cep.2022.01319.

341 Parker, W., Corrigan, P.T., Anderson, R., et al., "Evidence That Acetaminophen Triggers Autism in Susceptible Individuals Has Been Ignored and Mishandled for More Than a Decade," *Journal of the Academy of Public Health* (2025). https:// publichealth.realclearjournals.org/literature-syntheses/2025/10/evidence-that -acetaminophen-triggers-autism-in-susceptible-individuals-has-been-ignored-and -mishandled-for-more-than-a-decade/.
Jones, J.P., 3rd, Williamson, L., Konsoula, Z., Anderson, R., Reissner, K.J., and Parker, W., "Evaluating the Role of Susceptibility Inducing Cofactors and of Acetaminophen in the Etiology of Autism Spectrum Disorder," *Life (Basel)* 14, no. 8 (2024). https://doi.org/10.3390/life14080918.

342 Jones, J.P., 3rd, Williamson, L., Konsoula, Z., Anderson, R., Reissner, K.J., and Parker, W., "Evaluating the Role of Susceptibility Inducing Cofactors and of Acetaminophen in the Etiology of Autism Spectrum Disorder," *Life (Basel)* 14, no. 8 (2024). https://doi.org/10.3390/life14080918.

343 Editor, "Weaponizing Uncertainty in Science and in Public Health Puts People in Harm's Way," *Nature* 646, no. 8084 (2025): 259–60. https://doi.org/10.1038/ d41586-025-03167-5.
Pearson, H., and Ledford, H., "Trump Links Autism and Tylenol: Is There Any Truth to It?" *Nature* 646, no. 8083 (2025): 13–14. https://doi.org/10.1038/ d41586-025-02876-1.

Looi, M.K., and Bowie, K., "Autism: Trump Links Condition to Tylenol and Touts Leucovorin as 'First' US Therapeutic," *BMJ* 390 (2025): r2004. https://doi .org/10.1136/bmj.r2004.
Schweitzer, K., "Acetaminophen Use in Pregnancy-Study Author Explains the Data," *JAMA* 334, no. 17 (2025): 1499–501. https://doi.org/10.1001/ jama.2025.19345.
Andrade, C., "Maternal Use of Acetaminophen (Paracetamol) During Pregnancy and Neurodevelopmental Disorders in Offspring: A Reasoned Evaluation of Risk," *Journal of Clinical Psychiatry* 86, no. 4 (2025). https://doi.org/10.4088/ JCP.25f16187.
Louwen, F., Deuster, E., McAuliffe, F.M., et al., "Paracetamol (Acetaminophen) Use During Pregnancy and Autism Risk: Evidence Does Not Support Causal Association," *International Journal of Gynaecology & Obstetrics* 171, no. 3 (2025): 915–19. https://doi.org/10.1002/ijgo.70577.

344 Jones, J.P., 3rd, Williamson, L., Konsoula, Z., Anderson, R., Reissner, K.J., and Parker, W., "Evaluating the Role of Susceptibility Inducing Cofactors and of Acetaminophen in the Etiology of Autism Spectrum Disorder," *Life (Basel)* 14, no. 8 (2024). https://doi.org/10.3390/life14080918.

345 Baker, S., "An Interview with Dr. William Parker on the Connection between Acetaminophen and Autism," *Integrative Medicine (Encinitas)* 23, no. 5 (2024): 42–47.

346 Pearson, H., and Ledford, H., "Trump Links Autism and Tylenol: Is There Any Truth to It?" *Nature* 646, no. 8083 (2025): 13–14. https://doi.org/10.1038/ d41586-025-02876-1.

Chapter 11: Summary of Clinical Recommendations

347 Parker, W., Anderson, L.G., Jones, J.P., et al., "The Dangers of Acetaminophen for Neurodevelopment Outweigh Scant Evidence for Long-Term Benefits," *Children* 11, no. 1 (2024): 44. https://www.mdpi.com/2227-9067/11/1/44.

348 WHO, "Reducing Pain at the Time of Vaccination: WHO Position Paper, September 2015—Recommendations," *Vaccine* 34, no. 32 (2016): 3629–30. https://doi.org/10.1016/j.vaccine.2015.11.005.

349 Cendejas-Hernandez, J., Sarafian, J., Lawton, V., et al., "Paracetamol (Acetaminophen) Use in Infants and Children Was Never Shown to Be Safe for Neurodevelopment: A Systematic Review with Citation Tracking," *European Journal of Pediatrics* 181 (2022): 1835–57. https://doi.org/10.1007/ s00431-022-04407-w.

350 Cheng, A.M., Jaint, D., Thomas, S., Wilson, J., and Parker, W., "Overcoming Evolutionary Mismatch by Self-Treatment with Helminths: Current Practices and Experience," *Journal of Evolutionary Medicine* 3 (2015): Article ID 235910.
Liu, J., Morey, R.A., Wilson, J.K., and Parker, W., "Practices and Outcomes of Self-Treatment with Helminths Based on Physicians' Observations," *Journal of Helminthology* FirstView (2016): 1–11.
Venkatakrishnan, A., Sarafian, J.T., Jirků-Pomajbíková, K., and Parker, W., "Socio-Medical Studies of Individuals Self-Treating with Helminths Provide

Insight into Clinical Trial Design for Assessing Helminth Therapy," *Parasitology International* 87 (2021): 102488. https://doi.org/10.1016/j.parint.2021.102488.

351 Kim, J.H., and Scialli, A.R., "Thalidomide: The Tragedy of Birth Defects and the Effective Treatment of Disease," *Toxicological Sciences* 122, no. 1 (2011): 1–6. https://doi.org/10.1093/toxsci/kfr088.

352 Cendejas-Hernandez, J., Sarafian, J., Lawton, V., et al., "Paracetamol (Acetaminophen) Use in Infants and Children Was Never Shown to Be Safe for Neurodevelopment: A Systematic Review with Citation Tracking," *European Journal of Pediatrics* 181 (2022): 1835–57. https://doi.org/10.1007/s00431-022-04407-w.

353 Hart, B.L., "Biological Basis of the Behavior of Sick Animals," *Neuroscience & Biobehavioral Reviews* 12, no. 2 (1988): 123–37. http://www.ncbi.nlm.nih.gov/entrez/query.fcgi?cmd=Retrieve&db=PubMed&dopt=Citation&list_uids=3050629.
Evans, S.S., Repasky, E.A., and Fisher, D.T., "Fever and the Thermal Regulation of Immunity: The Immune System Feels the Heat," *Nature Reviews. Immunology* 15, no. 6 (2015): 335–49. https://doi.org/10.1038/nri3843.
Wrotek, S., LeGrand, E.K., Dzialuk, A., and Alcock, J., "Let Fever Do Its Job: The Meaning of Fever in the Pandemic Era," *Evolution, Medicine, and Public Health* 9, no. 1 (2021): 26–35. https://doi.org/10.1093/emph/eoaa044.

354 Homme, J.H., and Fischer, P.R., "Randomised Controlled Trial: Prophylactic Paracetamol at the Time of Infant Vaccination Reduces the Risk of Fever but Also Reduces Antibody Response," *Evidence-Based Medicine* 15, no. 2 (2010): 50–51. https://doi.org/10.1136/ebm1049.

355 Schmitt, B.D., "Fever Phobia: Misconceptions of Parents About Fevers," *American Journal of Diseases of Children* 134, no. 2 (1980): 176–81.
El-Radhi, A.S.M., "Fever Management: Evidence Vs Current Practice," *Clinical Pediatrics* 1 (2012): 29–33.
Sullivan, J.E., and Farrar, H.C., "Fever and Antipyretic Use in Children," *Pediatrics* 127, no. 3 (2011): 580–87. https://doi.org/10.1542/peds.2010-3852.

356 Ahlqvist, V.H., Sjöqvist, H., Dalman, C., et al., "Acetaminophen Use During Pregnancy and Children's Risk of Autism, ADHD, and Intellectual Disability," *JAMA* 331, no. 14 (2024): 1205–14. https://doi.org/10.1001/jama.2024.3172.

357 Jones, J.P., 3rd, Williamson, L., Konsoula, Z., Anderson, R., Reissner, K.J., and Parker, W., "Evaluating the Role of Susceptibility Inducing Cofactors and of Acetaminophen in the Etiology of Autism Spectrum Disorder," *Life (Basel)* 14, no. 8 (2024). https://doi.org/10.3390/life14080918.

358 Alberti, A., Pirrone, P., Elia, M., Waring, R.H., and Romano, C., "Sulphation Deficit in 'Low-Functioning' Autistic Children: A Pilot Study," *Biological Psychiatry* 46, no. 3 (1999): 420–24. https://doi.org/10.1016/s0006-3223(98)00337-0.

359 Geier, D.A., Kern, J.K., Garver, C.R., Adams, J.B., Audhya, T., and Geier, M.R., "A Prospective Study of Transsulfuration Biomarkers in Autistic Disorders," *Neurochemical Research* 34, no. 2 (2009): 386–93. https://doi.org/10.1007/s11064-008-9782-x.

Pagan, C., Benabou, M., Leblond, C., et al., "Decreased Phenol Sulfotransferase Activities Associated with Hyperserotonemia in Autism Spectrum Disorders," *Translational Psychiatry* 11, no. 1 (2021): 23. https://doi.org/10.1038/s41398-020-01125-5

360 Miller, R.P., Roberts, R.J., and Fischer, L.J., "Acetaminophen Elimination Kinetics in Neonates, Children, and Adults," *Clinical Pharmacology & Therapeutics* 19, no. 3 (1976): 284–94. https://doi.org/10.1002/cpt1976193284.
Cook, S.F., Stockmann, C., Samiee-Zafarghandy, S., et al., "Neonatal Maturation of Paracetamol (Acetaminophen) Glucuronidation, Sulfation, and Oxidation Based on a Parent-Metabolite Population Pharmacokinetic Model," *Clinical Pharmacokinetics* 55, no. 11 (2016): 1395–411. https://doi.org/10.1007/s40262-016-0408-1.

361 Stein, T.P., Schluter, M.D., Steer, R.A., and Ming, X., "Autism and Phthalate Metabolite Glucuronidation," *Journal of Autism and Developmental Disorders* 43, no. 11 (2013): 2677–85. https://doi.org/10.1007/s10803-013-1822-y.

362 Brzezinski, M.R., Boutelet-Bochan, H., Person, R.E., Fantel, A.G., and Juchau, M.R., "Catalytic Activity and Quantitation of Cytochrome P-450 2e1 in Prenatal Human Brain," *Journal of Pharmacology and Experimental Therapeutics* 289, no. 3 (1999): 1648–53.

363 Zhu, Y., Mordaunt, C.E., Yasui, D.H., et al., "Placental DNA Methylation Levels at CYP2E1 and IRS2 Are Associated with Child Outcome in a Prospective Autism Study," *Human Molecular Genetics* 28, no. 16 (2019): 2659–74. https://doi.org/10.1093/hmg/ddz084.

364 Santos, J.X., Rasga, C., Marques, A.R., et al., "A Role for Gene-Environment Interactions in Autism Spectrum Disorder Is Supported by Variants in Genes Regulating the Effects of Exposure to Xenobiotics," *Frontiers in Neuroscience* 16 (2022): 862315. https://doi.org/10.3389/fnins.2022.862315.
Braam, W., Keijzer, H., Struijker Boudier, H., Didden, R., Smits, M., and Curfs, L., "CYP1A2 Polymorphisms in Slow Melatonin Metabolisers: A Possible Relationship with Autism Spectrum Disorder?" *Journal of Intellectual Disability Research* 57, no. 11 (2013): 993–1000. https://doi.org/10.1111/j.1365-2788.2012.01595.x.

365 Fattorusso, A., Di Genova, L., Dell'Isola, G.B., Mencaroni, E., and Esposito, S., "Autism Spectrum Disorders and the Gut Microbiota," *Nutrients* 11, no. 3 (2019). https://doi.org/10.3390/nu11030521.
Peralta-Marzal, L.N., Prince, N., Bajic, D., et al., "The Impact of Gut Microbiota-Derived Metabolites in Autism Spectrum Disorders," *International Journal of Molecular Sciences* 22, no. 18 (2021). https://doi.org/10.3390/ijms221810052.

366 Chopyk, D.M., Stuart, J.D., Zimmerman, M.G., et al., "Acetaminophen Intoxication Rapidly Induces Apoptosis of Intestinal Crypt Stem Cells and Enhances Intestinal Permeability," *Hepatology Communications* 3, no. 11 (2019): 1435–49. https://doi.org/10.1002/hep4.1406.
Schneider, K.M., Elfers, C., Ghallab, A., et al., "Intestinal Dysbiosis Amplifies Acetaminophen-Induced Acute Liver Injury," *Cellular and Molecular Gastroenterology and Hepatology* 11, no. 4 (2021): 909–33.

Alabbas, S.Y., Giri, R., Oancea, I., et al., "Gut Inflammation and Adaptive Immunity Amplify Acetaminophen Toxicity in Bowel and Liver," *Journal of Gastroenterology and Hepatology* 38, no. 4 (2023): 609–18. https://doi.org/ https://doi.org/10.1111/jgh.16102.

367 McCrae, J.C., Morrison, E.E., MacIntyre, I.M., Dear, J.W., and Webb, D.J., "Long-Term Adverse Effects of Paracetamol—A Review," *British Journal of Clinical Pharmacology* 84, no. 10 (2018): 2218–30. https://doi.org/10.1111/ bcp.13656.

368 Parker, W., Hornik, C.D., Bilbo, S., et al., "The Role of Oxidative Stress, Inflammation and Acetaminophen Exposure from Birth to Early Childhood in the Induction of Autism," *Journal of International Medical Research* 45, no. 2 (2017): 407–38.

369 Flynn, C.K., Adams, J.B., Krajmalnik-Brown, R., et al., "Review of Elevated Para-Cresol in Autism and Possible Impact on Symptoms," *International Journal of Molecular Sciences* 26, no. 4 (2025). https://doi.org/10.3390/ijms26041513. Xiong, X., Liu, D., Wang, Y., Zeng, T., and Peng, Y., "Urinary 3-(3-Hydroxyphenyl)-3-Hydroxypropionic Acid, 3-Hydroxyphenylacetic Acid, and 3-Hydroxyhippuric Acid Are Elevated in Children with Autism Spectrum Disorders," *BioMed Research International* (2016): 9485412. https://doi. org/10.1155/2016/9485412.

370 Parker, W., Hornik, C.D., Bilbo, S., et al., "The Role of Oxidative Stress, Inflammation and Acetaminophen Exposure from Birth to Early Childhood in the Induction of Autism," *Journal of International Medical Research* 45, no. 2 (2017): 407–38. Alberti, A., Pirrone, P., Elia, M., Waring, R.H., and Romano, C., "Sulphation Deficit in 'Low-Functioning' Autistic Children: A Pilot Study," *Biological Psychiatry* 46, no. 3 (1999): 420–24. https://doi.org/10.1016/ s0006-3223(98)00337-0. Frye, R.E., Sequeira, J.M., Quadros, E.V., James, S.J., and Rossignol, D.A., "Cerebral Folate Receptor Autoantibodies in Autism Spectrum Disorder," *Molecular Psychiatry* 18, no. 3 (2013): 369–81. https://doi.org/10.1038/ mp.2011.175.

371 Hutabarat, R.M., Unadkat, J.D., Kushmerick, P., Aitken, M.L., Slattery, J.T., and Smith, A.L., "Disposition of Drugs in Cystic Fibrosis. III. Acetaminophen," *Clinical Pharmacology & Therapeutics* 50, no. 6 (1991): 695–701. Kearns, G.L., "Hepatic Drug Metabolism in Cystic Fibrosis: Recent Developments and Future Directions," *Annals of Pharmacotherapy* 27, no. 1 (1993): 74–79.

372 Parker, W., Hornik, C.D., Bilbo, S., et al., "The Role of Oxidative Stress, Inflammation and Acetaminophen Exposure from Birth to Early Childhood in the Induction of Autism," *Journal of International Medical Research* 45, no. 2 (2017): 407–38.

373 Pourtavakoli, A., and Ghafouri-Fard, S., "Calcium Signaling in Neurodevelopment and Pathophysiology of Autism Spectrum Disorders," *Molecular Biology Reports* 49, no. 11 (2022): 10811–23. https://doi.org/10.1007/ s11033-022-07775-6.

374 Salas, V.M., and Corcoran, G.B., "Calcium-Dependent DNA Damage and Adenosine 3',5'-Cyclic Monophosphate-Independent Glycogen Phosphorylase Activation in an in Vitro Model of Acetaminophen-Induced Liver Injury," *Hepatology* 25, no. 6 (1997): 1432–38. https://doi.org/10.1002/hep.510250621.

375 Schmunk, G., Nguyen, R.L., Ferguson, D.L., Kumar, K., Parker, I., and Gargus, J.J., "High-Throughput Screen Detects Calcium Signaling Dysfunction in Typical Sporadic Autism Spectrum Disorder," *Scientific Reports* 7 (2017): 40740. https://doi.org/10.1038/srep40740.

376 Hogestatt, E.D., Jonsson, B.A., Ermund, A., et al., "Conversion of Acetaminophen to the Bioactive N-Acylphenolamine AM_{404} Via Fatty Acid Amide Hydrolase-Dependent Arachidonic Acid Conjugation in the Nervous System," *Journal of Biological Chemistry* 280, no. 36 (2005): 31405–12. https://doi.org/10.1074/jbc.M501489200.

377 Hinz, B., Cheremina, O., and Brune, K., "Acetaminophen (Paracetamol) Is a Selective Cyclooxygenase-2 Inhibitor in Man," *FASEB Journal* 22, no. 2 (2008): 383–90. https://doi.org/10.1096/fj.07-8506com.

378 Hirai, T., Umeda, N., Harada, T., et al., "Arachidonic Acid-Derived Dihydroxy Fatty Acids in Neonatal Cord Blood Relate Symptoms of Autism Spectrum Disorders and Social Adaptive Functioning: Hamamatsu Birth Cohort for Mothers and Children (HBC Study)," *Psychiatry and Clinical Neurosciences* 78, no. 9 (2024): 546–57. https://doi.org/10.1111/pcn.13710.

379 El-Ansary, A., Alfawaz, H.A., Bacha, A.B., and Al-Ayadhi, L.Y., "Combining Anti-Mitochondrial Antibodies, Anti-Histone, and Pla2/Cox Biomarkers to Increase Their Diagnostic Accuracy for Autism Spectrum Disorders," *Brain Sciences* 14, no. 6 (2024): 576. https://www.mdpi.com/2076-3425/14/6/576.

380 Schultz, S.T., Klonoff-Cohen, H.S., Wingard, D.L., Akshoomoff, N.A., Macera, C.A., and Ji, M. "Acetaminophen (Paracetamol) Use, Measles-Mumps-Rubella Vaccination, and Autistic Disorder. The Results of a Parent Survey," *Autism* 12, no. 3 (2008): 293–307.

381 Zhao, L., Jones, J., Anderson, L., et al., "Acetaminophen Causes Neurodevelopmental Injury in Susceptible Babies and Children: No Valid Rationale for Controversy," *Clinical and Experimental Pediatrics* (2023). https://doi.org/10.3345/cep.2022.01319.

382 Schultz, S.T., Klonoff-Cohen, H.S., Wingard, D.L., Akshoomoff, N.A., Macera, C.A., and Ji, M. "Acetaminophen (Paracetamol) Use, Measles-Mumps-Rubella Vaccination, and Autistic Disorder. The Results of a Parent Survey," *Autism* 12, no. 3 (2008): 293–307.

383 Ji, Y., Azuine, R.E., Zhang, Y., et al., "Association of Cord Plasma Biomarkers of in Utero Acetaminophen Exposure with Risk of Attention-Deficit/Hyperactivity Disorder and Autism Spectrum Disorder in Childhood," *JAMA Psychiatry* 77, no. 2 (2020): 180–89. https://doi.org/10.1001/jamapsychiatry.2019.3259.

384 Ji, Y., Azuine, R.E., Zhang, Y., et al., "Association of Cord Plasma Biomarkers of in Utero Acetaminophen Exposure with Risk of Attention-Deficit/Hyperactivity Disorder and Autism Spectrum Disorder in Childhood," *JAMA Psychiatry* 77, no. 2 (2020): 180–89. https://doi.org/10.1001/jamapsychiatry.2019.3259.

Tovo-Rodrigues, L., Schneider, B.C., Martins-Silva, T., et al., "Is Intrauterine Exposure to Acetaminophen Associated with Emotional and Hyperactivity Problems During Childhood? Findings from the 2004 Pelotas Birth Cohort," *BMC Psychiatry* 18, no. 1 (2018): 368. https://doi.org/10.1186/s12888-018-1942-1.

Vlenterie, R., Wood, M.E., Brandlistuen, R.E., Roeleveld, N., van Gelder, M.M., and Nordeng, H., "Neurodevelopmental Problems at 18 Months among Children Exposed to Paracetamol in Utero: A Propensity Score Matched Cohort Study," *International Journal of Epidemiology* 45, no. 6 (2016): 1998–2008. https://doi.org/10.1093/ije/dyw192.

Liew, Z., Ritz, B., Virk, J., Arah, O.A., and Olsen, J., "Prenatal Use of Acetaminophen and Child IQ: A Danish Cohort Study," *Epidemiology* 27, no. 6 (2016): 912–18. https://doi.org/10.1097/ede.0000000000000540.

Liew, Z., Bach, C.C., Asarnow, R.F., Ritz, B., and Olsen, J., "Paracetamol Use During Pregnancy and Attention and Executive Function in Offspring at Age 5 Years," *International Journal of Epidemiology* 45, no. 6 (2016): 2009–17. https://doi.org/10.1093/ije/dyw296.

Avella-Garcia, C.B., Julvez, J., Fortuny, J., et al., "Acetaminophen Use in Pregnancy and Neurodevelopment: Attention Function and Autism Spectrum Symptoms," *International Journal of Epidemiology* 45, no. 6 (2016): 1987–96. https://doi.org/10.1093/ije/dyw115.

Alemany, S., Avella-García, C., Liew, Z., et al., "Prenatal and Postnatal Exposure to Acetaminophen in Relation to Autism Spectrum and Attention-Deficit and Hyperactivity Symptoms in Childhood: Meta-Analysis in Six European Population-Based Cohorts," *European Journal of Epidemiology* 36, no. 10 (2021): 993–1004. https://doi.org/10.1007/s10654-021-00754-4.

Skovlund, E., Handal, M., Selmer, R., Brandlistuen, R.E., and Skurtveit, S., "Language Competence and Communication Skills in 3-Year-Old Children after Prenatal Exposure to Analgesic Opioids," *Pharmacoepidemiology and Drug Safety* 26, no. 6 (2017): 625–34. https://doi.org/10.1002/pds.4170.

Liew, Z., Ritz, B., Rebordosa, C., Lee, P.C., and Olsen, J., "Acetaminophen Use During Pregnancy, Behavioral Problems, and Hyperkinetic Disorders," *JAMA Pediatrics* 168, no. 4 (2014): 313–20. https://doi.org/10.1001/jamapediatrics.2013.4914.

Liew, Z., Ritz, B., Virk, J., and Olsen, J., "Maternal Use of Acetaminophen During Pregnancy and Risk of Autism Spectrum Disorders in Childhood: A Danish National Birth Cohort Study," *Autism Research* 9, no. 9 (2016): 951–58. https://doi.org/10.1002/aur.1591.

Ystrom, E., Gustavson, K., Brandlistuen, R.E., et al., "Prenatal Exposure to Acetaminophen and Risk of ADHD," *Pediatrics* 140, no. 5 (2017). https://doi.org/10.1542/peds.2016-3840.

Thompson, J.M., Waldie, K.E., Wall, C.R., Murphy, R., and Mitchell, E.A., "Associations between Acetaminophen Use During Pregnancy and ADHD Symptoms Measured at Ages 7 and 11 Years," *PLOS One* 9, no. 9 (2014): e108210. https://doi.org/10.1371/journal.pone.0108210.

Stergiakouli, E., Thapar, A., and Davey Smith, G., "Association of Acetaminophen Use During Pregnancy with Behavioral Problems in Childhood: Evidence against Confounding," *JAMA Pediatrics* 170, no. 10 (2016): 964–70. https://doi.org/10.1001/jamapediatrics.2016.1775.

Brandlistuen, R.E., Ystrom, E., Nulman, I., Koren, G., and Nordeng, H., "Prenatal Paracetamol Exposure and Child Neurodevelopment: A Sibling-Controlled Cohort Study," *International Journal of Epidemiology* 42, no. 6 (2013): 1702–13.

Sznajder, K.K., Teti, D.M., and Kjerulff, K.H., "Maternal Use of Acetaminophen During Pregnancy and Neurobehavioral Problems in Offspring at 3 Years: A Prospective Cohort Study," *PLOS One* 17, no. 9 (2022): e0272593. https://doi.org/10.1371/journal.pone.0272593.

Inoue, K., Ritz, B., Ernst, A., et al., "Behavioral Problems at Age 11 Years after Prenatal and Postnatal Exposure to Acetaminophen: Parent-Reported and Self-Reported Outcomes," *American Journal of Epidemiology* 190, no. 6 (2021): 1009–20. https://doi.org/10.1093/aje/kwaa257.

Liew, Z., Kioumourtzoglou, M.A., Roberts, A.L., O'Reilly É, J., Ascherio, A., and Weisskopf, M.G., "Use of Negative Control Exposure Analysis to Evaluate Confounding: An Example of Acetaminophen Exposure and Attention-Deficit/Hyperactivity Disorder in Nurses' Health Study II," *American Journal of Epidemiology* 188, no. 4 (2019): 768–75. https://doi.org/10.1093/aje/kwy288.

Parker, S.E., Collett, B.R., and Werler, M.M., "Maternal Acetaminophen Use During Pregnancy and Childhood Behavioural Problems: Discrepancies between Mother- and Teacher-Reported Outcomes," *Paediatric and Perinatal Epidemiology* 34, no. 3 (2020): 299–308. https://doi.org/10.1111/ppe.12601.

Trønnes, J.N., Wood, M., Lupattelli, A., Ystrom, E., and Nordeng, H., "Prenatal Paracetamol Exposure and Neurodevelopmental Outcomes in Preschool-Aged Children," *Paediatric and Perinatal Epidemiology* 34, no. 3 (2020): 247–56. https://doi.org/10.1111/ppe.12568.

Golding, J., Gregory, S., Clark, R., Ellis, G., Iles-Caven, Y., and Northstone, K., "Associations between Paracetamol (Acetaminophen) Intake between 18 and 32 Weeks Gestation and Neurocognitive Outcomes in the Child: A Longitudinal Cohort Study," *Paediatric and Perinatal Epidemiology* 34, no. 3 (2020): 257–66. https://doi.org/10.1111/ppe.12582.

Bornehag, C.G., Reichenberg, A., Hallerback, M.U., et al., "Prenatal Exposure to Acetaminophen and Children's Language Development at 30 Months," *European Psychiatry* 51 (2018): 98–103. https://doi.org/10.1016/j.eurpsy.2017.10.007.

Chen, M.H., Pan, T.L., Wang, P.W., et al. "Prenatal Exposure to Acetaminophen and the Risk of Attention-Deficit/Hyperactivity Disorder: A Nationwide Study in Taiwan," *Journal of Clinical Psychiatry* 80, no. 5 (2019). https://doi.org/10.4088/JCP.18m12612.

Baker, B.H., Lugo-Candelas, C., Wu, H., et al., "Association of Prenatal Acetaminophen Exposure Measured in Meconium with Risk of Attention-Deficit/Hyperactivity Disorder Mediated by Frontoparietal Network Brain Connectivity," *JAMA Pediatrics* 174, no. 11 (2020): 1073–81. https://doi.org/10.1001/jamapediatrics.2020.3080.

Baker, B.H., Bammler, T.K., Barrett, E.S., et al., "Associations of Maternal Blood Biomarkers of Prenatal Apap Exposure with Placental Gene Expression and Child Attention Deficit Hyperactivity Disorder," *Nature Mental Health* (2025). https://doi.org/10.1038/s44220-025-00387-6.

385 Jones, J.P., 3rd, Williamson, L., Konsoula, Z., Anderson, R., Reissner, K.J., and Parker, W., "Evaluating the Role of Susceptibility Inducing Cofactors and of Acetaminophen in the Etiology of Autism Spectrum Disorder," *Life (Basel)* 14, no. 8 (2024). https://doi.org/10.3390/life14080918.

386 Ahlqvist, V.H., Sjöqvist, H., Dalman, C., et al., "Acetaminophen Use During Pregnancy and Children's Risk of Autism, ADHD, and Intellectual Disability," *JAMA* 331, no. 14 (2024): 1205–14. https://doi.org/10.1001/jama.2024.3172.

387 Patel, E., Jones, J.P., 3rd, Bono-Lunn, D., et al., "The Safety of Pediatric Use of Paracetamol (Acetaminophen): A Narrative Review of Direct and Indirect Evidence," *Minerva Pediatrics (Torino)* 74, no. 6 (2022): 774–88. https://doi.org/10.23736/s2724-5276.22.06932-4.

Jones, J.P., 3rd, Williamson, L., Konsoula, Z., Anderson, R., Reissner, K.J., and Parker, W., "Evaluating the Role of Susceptibility Inducing Cofactors and of Acetaminophen in the Etiology of Autism Spectrum Disorder," *Life (Basel)* 14, no. 8 (2024). https://doi.org/10.3390/life14080918.

Baker, B.H., Bammler, T.K., Barrett, E.S., et al., "Associations of Maternal Blood Biomarkers of Prenatal Apap Exposure with Placental Gene Expression and Child Attention Deficit Hyperactivity Disorder," *Nature Mental Health* (2025). https://doi.org/10.1038/s44220-025-00387-6.

388 Alemany, S., Avella-García, C., Liew, Z., et al., "Prenatal and Postnatal Exposure to Acetaminophen in Relation to Autism Spectrum and Attention-Deficit and Hyperactivity Symptoms in Childhood: Meta-Analysis in Six European Population-Based Cohorts," *European Journal of Epidemiology* 36, no. 10 (2021): 993–1004. https://doi.org/10.1007/s10654-021-00754-4.

389 Zhao, L., Jones, J., Anderson, L., et al., "Acetaminophen Causes Neurodevelopmental Injury in Susceptible Babies and Children: No Valid Rationale for Controversy," *Clinical and Experimental Pediatrics* (2023). https://doi.org/10.3345/cep.2022.01319.

390 Alemany, S., Avella-García, C., Liew, Z., et al., "Prenatal and Postnatal Exposure to Acetaminophen in Relation to Autism Spectrum and Attention-Deficit and Hyperactivity Symptoms in Childhood: Meta-Analysis in Six European Population-Based Cohorts," *European Journal of Epidemiology* 36, no. 10 (2021): 993–1004. https://doi.org/10.1007/s10654-021-00754-4.

Zhao, L., Jones, J., Anderson, L., et al., "Acetaminophen Causes Neurodevelopmental Injury in Susceptible Babies and Children: No Valid Rationale for Controversy," *Clinical and Experimental Pediatrics* (2023). https://doi.org/10.3345/cep.2022.01319.

391 Patel, E., Jones, J.P., 3rd, Bono-Lunn, D., et al., "The Safety of Pediatric Use of Paracetamol (Acetaminophen): A Narrative Review of Direct and Indirect Evidence," *Minerva Pediatrics (Torino)* 74, no. 6 (2022): 774–88. https://doi.org/10.23736/s2724-5276.22.06932-4.

392 Graeca, M., and Kulesza, R., "Impaired Brainstem Auditory Evoked Potentials after in Utero Exposure to High Dose Paracetamol Exposure," *Hearing Research* 454 (2024): 109149. https://doi.org/10.1016/j.heares.2024.109149.

393 Alemany, S., Avella-García, C., Liew, Z., et al., "Prenatal and Postnatal Exposure to Acetaminophen in Relation to Autism Spectrum and Attention-Deficit and Hyperactivity Symptoms in Childhood: Meta-Analysis in Six European Population-Based Cohorts," *European Journal of Epidemiology* 36, no. 10 (2021): 993–1004. https://doi.org/10.1007/s10654-021-00754-4.

394 Alemany, S., Avella-García, C., Liew, Z., et al., "Prenatal and Postnatal Exposure to Acetaminophen in Relation to Autism Spectrum and Attention-Deficit and Hyperactivity Symptoms in Childhood: Meta-Analysis in Six European Population-Based Cohorts," *European Journal of Epidemiology* 36, no. 10 (2021): 993–1004. https://doi.org/10.1007/s10654-021-00754-4.
Jones, J.P., 3rd, Williamson, L., Konsoula, Z., Anderson, R., Reissner, K.J., and Parker, W., "Evaluating the Role of Susceptibility Inducing Cofactors and of Acetaminophen in the Etiology of Autism Spectrum Disorder," *Life (Basel)* 14, no. 8 (2024). https://doi.org/10.3390/life14080918.

395 Parker, W., Hornik, C.D., Bilbo, S., et al., "The Role of Oxidative Stress, Inflammation and Acetaminophen Exposure from Birth to Early Childhood in the Induction of Autism," *Journal of International Medical Research* 45, no. 2 (2017): 407–38.

396 Zhao, L., Jones, J., Anderson, L., et al., "Acetaminophen Causes Neurodevelopmental Injury in Susceptible Babies and Children: No Valid Rationale for Controversy," *Clinical and Experimental Pediatrics* (2023). https://doi.org/10.3345/cep.2022.01319.
Jones, J.P., 3rd, Williamson, L., Konsoula, Z., Anderson, R., Reissner, K.J., and Parker, W., "Evaluating the Role of Susceptibility Inducing Cofactors and of Acetaminophen in the Etiology of Autism Spectrum Disorder," *Life (Basel)* 14, no. 8 (2024). https://doi.org/10.3390/life14080918.

397 Parker, W., Hornik, C.D., Bilbo, S., et al., "The Role of Oxidative Stress, Inflammation and Acetaminophen Exposure from Birth to Early Childhood in the Induction of Autism," *Journal of International Medical Research* 45, no. 2 (2017): 407–38.

398 Donohue, J., "A History of Drug Advertising: The Evolving Roles of Consumers and Consumer Protection," *The Milbank Quarterly* 84, no. 4 (2006): 659–99. https://doi.org/10.1111/j.1468-0009.2006.00464.x.

399 Zhao, L., Jones, J., Anderson, L., et al., "Acetaminophen Causes Neurodevelopmental Injury in Susceptible Babies and Children: No Valid Rationale for Controversy," *Clinical and Experimental Pediatrics* (2023). https://doi.org/10.3345/cep.2022.01319.
Jones, J.P., 3rd, Williamson, L., Konsoula, Z., Anderson, R., Reissner, K.J., and Parker, W., "Evaluating the Role of Susceptibility Inducing Cofactors and of Acetaminophen in the Etiology of Autism Spectrum Disorder," *Life (Basel)* 14, no. 8 (2024). https://doi.org/10.3390/life14080918.

400 Rimland, B., "The Autism Increase: Research Needed on the Vaccine Connection," *Autism Research Review International* 14, no. 1 (2000): 3.

401 Parker, W., Hornik, C.D., Bilbo, S., et al., "The Role of Oxidative Stress, Inflammation and Acetaminophen Exposure from Birth to Early Childhood in the Induction of Autism," *Journal of International Medical Research* 45, no. 2 (2017): 407–38.

402 Frisch, M., and Simonsen, J., "Ritual Circumcision and Risk of Autism Spectrum Disorder in 0- to 9-Year-Old Boys: National Cohort Study in Denmark," *Journal of the Royal Society of Medicine* 108, no. 7 (2015): 266–79. https://doi.org/10.1177/0141076814565942.

403 Frisch, M., and Simonsen, J., "Ritual Circumcision and Risk of Autism Spectrum Disorder in 0- to 9-Year-Old Boys: National Cohort Study in Denmark," *Journal of the Royal Society of Medicine* 108, no. 7 (2015): 266–79. https://doi.org/10.1177/0141076814565942.

404 Miani, A., Di Bernardo, G.A., Højgaard, A.D., et al. "Neonatal Male Circumcision Is Associated with Altered Adult Socio-Affective Processing," *Heliyon* 6, no. 11 (2020): e05566. https://doi.org/10.1016/j.heliyon.2020.e05566.

405 Howard, C.R., Howard, F.M., and Weitzman, M.L., "Acetaminophen Analgesia in Neonatal Circumcision: The Effect on Pain," *Pediatrics* 93, no. 4 (1994): 641–46.

406 Parker, W., Anderson, L.G., Jones, J.P., et al., "The Dangers of Acetaminophen for Neurodevelopment Outweigh Scant Evidence for Long-Term Benefits," *Children* 11, no. 1 (2024): 44. https://www.mdpi.com/2227-9067/11/1/44.

407 Raz, R., Weisskopf, M.G., Davidovitch, M., Pinto, O., and Levine, H., "Differences in Autism Spectrum Disorders Incidence by Sub-Populations in Israel 1992–2009: A Total Population Study," *Journal of Autism and Developmental Disorders* 45, no. 4 (2015): 1062–69. https://doi.org/10.1007/s10803-014-2262-z.

408 Zauderer, C., "Maternity Care for Orthodox Jewish Couples: Implications for Nurses in the Obstetric Setting," *Nursing for Women's Health* 13, no. 2 (2009): 112–20. https://doi.org/10.1111/j.1751-486X.2009.01402.x.

409 Howard, C.R., Howard, F.M., and Weitzman, M.L., "Acetaminophen Analgesia in Neonatal Circumcision: The Effect on Pain," *Pediatrics* 93, no. 4 (1994): 641–46.

410 Kacker, S., and Tobian, A.A., "Male Circumcision: Integrating Tradition and Medical Evidence," *Israel Medical Association Journal* 15, no. 1 (2013): 37–38.

411 Zauderer, C., "Maternity Care for Orthodox Jewish Couples: Implications for Nurses in the Obstetric Setting," *Nursing for Women's Health* 13, no. 2 (2009): 112–20. https://doi.org/10.1111/j.1751-486X.2009.01402.x.

412 Parker, W., Anderson, L.G., Jones, J.P., et al., "The Dangers of Acetaminophen for Neurodevelopment Outweigh Scant Evidence for Long-Term Benefits," *Children* 11, no. 1 (2024): 44. https://www.mdpi.com/2227-9067/11/1/44.

413 Jones, J.P., 3rd, Williamson, L., Konsoula, Z., Anderson, R., Reissner, K.J., and Parker, W., "Evaluating the Role of Susceptibility Inducing Cofactors and of Acetaminophen in the Etiology of Autism Spectrum Disorder," *Life (Basel)* 14, no. 8 (2024). https://doi.org/10.3390/life14080918.

414 Jones, J.P., 3rd, Williamson, L., Konsoula, Z., Anderson, R., Reissner, K.J., and Parker, W., "Evaluating the Role of Susceptibility Inducing Cofactors and of Acetaminophen in the Etiology of Autism Spectrum Disorder," *Life (Basel)* 14, no. 8 (2024). https://doi.org/10.3390/life14080918.

415 Aguiar, A.G., Mainegra, F.D., García, R.O., and Hernandez, F.Y., "Diagnosis in Children with Autism Spectrum Disorders in Their Development in Textual Comprehension," *Revista de Ciencias Medicas de Pinar del Río* 20, no. 6 (2016): 729–37.

416 Pharmacy Times, "A Peek at Pharmacy Practice in Cuba," May 24, 2016, https://www.pharmacytimes.com/view/a-peek-at-pharmacy-practice-in-cuba.

417 C. Yeldham, "Going to Cuba? Here's What Else to Pack," STARTUP CUBA.TV., Nov. 10, 2021, https://startupcuba.tv/2021/11/10/going-to-cuba-heres-what-else-to-pack/.
B. Sainsbury, "20 Things to Know before Visiting Cuba," Lonely Planet, Dec. 23, 2025, https://www.lonelyplanet.com/articles/things-to-know-before-traveling-to-cuba.
Simply Cuba Tours, "Preparing for Cuba: The Ultimate Packing List for Health & Safety," Dec. 11, 2025, https://simplycubatours.com/preparing-for-cuba-the-ultimate-packing-list-for-health-safety/.
Locally Sourced Cuba Tours, "How to Support Local Cubans When You Travel to Cuba," accessed Feb. 7, 2026, https://locallysourcedcuba.com/how-to-support-locals-in-cuba/.

418 Kamer, A., Zohar, A.H., Youngmann, R., Diamond, G.W., Inbar, D., and Senecky, Y., "A Prevalence Estimate of Pervasive Developmental Disorder among Immigrants to Israel and Israeli Natives—A File Review Study," *Social Psychiatry and Psychiatric Epidemiology* 39, no. 2 (2004): 141–45. https://doi.org/10.1007/s00127-004-0696-x.

419 Suda, N., Hernandez, J.C., Poulton, J., et al., "Therapeutic Doses of Paracetamol with Co-Administration of Cysteine and Mannitol During Early Development Result in Long Term Behavioral Changes in Laboratory Rats," *PLOS One* 16, no. 6 (2020): e0253543. https://doi.org/https://doi.org/10.1371/journal.pone.0253543.
Viberg, H., Eriksson, P., Gordh, T., and Fredriksson, A., "Paracetamol (Acetaminophen) Administration During Neonatal Brain Development Affects Cognitive Function and Alters Its Analgesic and Anxiolytic Response in Adult Male Mice," *Toxicological Sciences* 138, no. 1 (2013): 139–47. https://doi.org/10.1093/toxsci/kft329.
Graeca, M., and Kulesza, R., "Impaired Brainstem Auditory Evoked Potentials after in Utero Exposure to High Dose Paracetamol Exposure," *Hearing Research* 454 (2024): 109149. https://doi.org/10.1016/j.heares.2024.109149.
Hussin, S., and Al-Allaf, L., "Histological Changes of Ca and Dg Regions of Hippocampus of Rats' Brain after Exposure to Acetaminophen in Postnatal Period," *Iraqi Journal of Veterinary Sciences* 36, no. 1 (2022): 151–58.
Harshaw, C., and Warner, A.G., "Interleukin-1β-Induced Inflammation and Acetaminophen During Infancy: Distinct and Interactive Effects on Social-Emotional and Repetitive Behavior in C57BL/6J Mice," *Pharmacology*

Biochemistry and Behavior 220 (2022): 173463. https://doi.org/10.1016/
j.pbb.2022.173463.

Herrington, J.A., Guss Darwich, J., Harshaw, C., Brigande, A.M., Leif, E.B.,
and Currie, P.J., "Elevated Ghrelin Alters the Behavioral Effects of Perinatal
Acetaminophen Exposure in Rats," *Developmental Psychobiology* 64, no. 3 (2022):
e22252. https://doi.org/10.1002/dev.22252.

Blecharz-Klin, K., Wawer, A., Jawna-Zboińska, K., et al., "Early Paracetamol
Exposure Decreases Brain-Derived Neurotrophic Factor (Bdnf) in Striatum and
Affects Social Behaviour and Exploration in Rats," *Pharmacology Biochemistry and
Behavior* 168 (2018): 25–32. https://doi.org/10.1016/j.pbb.2018.03.004.

Hay-Schmidt, A., Finkielman, O.T.E., Jensen, B.A.H., et al., "Prenatal Exposure
to Paracetamol/Acetaminophen and Precursor Aniline Impairs Masculinisation of
Male Brain and Behaviour," *Reproduction* 154, no. 2 (2017): 145–52. https://doi
.org/10.1530/rep-17-0165.

Baker, B.H., Rafikian, E.E., Hamblin, P.B., Strait, M.D., Yang, M., and
Pearson, B.L., "Sex-Specific Neurobehavioral and Prefrontal Cortex Gene
Expression Alterations Following Developmental Acetaminophen Exposure in
Mice," *Neurobiology of Disease* 177 (2023): 105970. https://doi.org/10.1016/
j.nbd.2022.105970.

Saad, A., Hegde, S., Kechichian, T., et al., "Is There a Causal Relation between
Maternal Acetaminophen Administration and Adhd?" *PLOS One* 11, no. 6
(2016): e0157380. https://doi.org/10.1371/journal.pone.0157380.

Klein, R.M., Motomura, V.N., Debiasi, J.D., and Moreira, E.G., "Gestational
Paracetamol Exposure Induces Core Behaviors of Neurodevelopmental Disorders
in Infant Rats and Modifies Response to a Cannabinoid Agonist in Females,"
Neurotoxicology and Teratology 99 (2023): 107279. https://doi.org/10.1016/
j.ntt.2023.107279.

Klein, R.M., Rigobello, C., Vidigal, C.B., et al., "Gestational Exposure to
Paracetamol in Rats Induces Neurofunctional Alterations in the Progeny,"
Neurotoxicology and Teratology 77 (2020): 106838. https://doi.org/10.1016/
j.ntt.2019.106838.

Philippot, G., Gordh, T., Fredriksson, A., and Viberg, H., "Adult
Neurobehavioral Alterations in Male and Female Mice Following Developmental
Exposure to Paracetamol (Acetaminophen): Characterization of a Critical
Period," *Journal of Applied Toxicology* 37, no. 10 (2017): 1174–81. https://doi
.org/10.1002/jat.3473.

Dean, S.L., Knutson, J.F., Krebs-Kraft, D.L., and McCarthy, M.M.,
"Prostaglandin E2 Is an Endogenous Modulator of Cerebellar Development
and Complex Behavior During a Sensitive Postnatal Period," *European Journal
of Neuroscience* 35, no. 8 (2012): 1218–29. https://doi.org/10.1111/j
.1460-9568.2012.08032.x.

Philippot, G., Hosseini, K., Yakub, A., et al., "Paracetamol (Acetaminophen)
and Its Effect on the Developing Mouse Brain," *Frontiers in Toxicology* 4 (2022):
867748. https://doi.org/10.3389/ftox.2022.867748.

420 Viberg, H., Eriksson, P., Gordh, T., and Fredriksson, A., "Paracetamol
(Acetaminophen) Administration During Neonatal Brain Development Affects

Cognitive Function and Alters Its Analgesic and Anxiolytic Response in Adult Male Mice," *Toxicological Sciences* 138, no. 1 (2013): 139–47. https://doi.org/10.1093/toxsci/kft329.

421 Suda, N., Hernandez, J.C., Poulton, J., et al., "Therapeutic Doses of Paracetamol with Co-Administration of Cysteine and Mannitol During Early Development Result in Long Term Behavioral Changes in Laboratory Rats," *PLOS One* 16, no.6 (2020): e0253543. https://doi.org/https://doi.org/10.1371/journal.pone.0253543.

422 Baker, B.H., Rafikian, E.E., Hamblin, P.B., Strait, M.D., Yang, M., and Pearson, B.L., "Sex-Specific Neurobehavioral and Prefrontal Cortex Gene Expression Alterations Following Developmental Acetaminophen Exposure in Mice," *Neurobiology of Disease* 177 (2023): 105970. https://doi.org/10.1016/j.nbd.2022.105970.
Dean, S.L., Knutson, J.F., Krebs-Kraft, D.L., and McCarthy, M.M., "Prostaglandin E2 Is an Endogenous Modulator of Cerebellar Development and Complex Behavior During a Sensitive Postnatal Period," *European Journal of Neuroscience* 35, no. 8 (2012): 1218–29. https://doi.org/10.1111/j.1460-9568.2012.08032.x.
Kanno, S.I., Tomizawa, A., Yomogida, S., and Hara, A., "Glutathione Peroxidase 3 Is a Protective Factor against Acetaminophen-Induced Hepatotoxicity in Vivo and in Vitro," *International Journal of Molecular Medicine* 40, no. 3 (2017): 748–54. https://doi.org/10.3892/ijmm.2017.3049.
Rigobello, C., Klein, R.M., Debiasi, J.D., et al., "Perinatal Exposure to Paracetamol: Dose and Sex-Dependent Effects in Behaviour and Brain's Oxidative Stress Markers in Progeny," *Behavioural Brain Research* 408 (2021): 113294. https://doi.org/10.1016/j.bbr.2021.113294.

423 Dean, S.L., Knutson, J.F., Krebs-Kraft, D.L., and McCarthy, M.M., "Prostaglandin E2 Is an Endogenous Modulator of Cerebellar Development and Complex Behavior During a Sensitive Postnatal Period," *European Journal of Neuroscience* 35, no. 8 (2012): 1218–29. https://doi.org/10.1111/j.1460-9568.2012.08032.x.
Kanno, S.I., Tomizawa, A., Yomogida, S., and Hara, A., "Glutathione Peroxidase 3 Is a Protective Factor against Acetaminophen-Induced Hepatotoxicity in Vivo and in Vitro," *International Journal of Molecular Medicine* 40, no. 3 (2017): 748–54. https://doi.org/10.3892/ijmm.2017.3049.

424 Graeca, M., and Kulesza, R., "Impaired Brainstem Auditory Evoked Potentials after in Utero Exposure to High Dose Paracetamol Exposure," *Hearing Research* 454 (2024): 109149. https://doi.org/10.1016/j.heares.2024.109149.

425 Klein, R.M., Rigobello, C., Vidigal, C.B., et al., "Gestational Exposure to Paracetamol in Rats Induces Neurofunctional Alterations in the Progeny," *Neurotoxicology and Teratology* 77 (2020): 106838. https://doi.org/10.1016/j.ntt.2019.106838.

426 Xu, M., Minagawa, Y., Kumazaki, H., Okada, K.I., and Naoi, N., "Prefrontal Responses to Odors in Individuals with Autism Spectrum Disorders: Functional NIRS Measurement Combined with a Fragrance Pulse Ejection System,"

Frontiers in Human Neuroscience 14 (2020): 523456. https://doi.org/10.3389/fnhum.2020.523456.

427 Graeca, M., and Kulesza, R., "Impaired Brainstem Auditory Evoked Potentials after in Utero Exposure to High Dose Paracetamol Exposure," *Hearing Research* 454 (2024): 109149. https://doi.org/10.1016/j.heares.2024.109149.

428 Posadas, I., Santos, P., Blanco, A., Muñoz-Fernández, M., and Ceña, V., "Acetaminophen Induces Apoptosis in Rat Cortical Neurons," *PLOS One* 5, no. 12 (2010): e15360. https://doi.org/10.1371/journal.pone.0015360.

429 Donovan, A.P., and Basson, M.A., "The Neuroanatomy of Autism—A Developmental Perspective," *Journal of Anatomy* 230, no. 1 (2017): 4–15. https://doi.org/10.1111/joa.12542.
Casanova, M.F., Sokhadze, E.M., Casanova, E.L., et al., "Translational Neuroscience in Autism: From Neuropathology to Transcranial Magnetic Stimulation Therapies," *Psychiatric Clinics of North Am* 43, no. 2 (2020): 229–48. https://doi.org/10.1016/j.psc.2020.02.004.

430 Dong, D., Zielke, H.R., Yeh, D., and Yang, P., "Cellular Stress and Apoptosis Contribute to the Pathogenesis of Autism Spectrum Disorder," *Autism Research* 11, no. 7 (2018): 1076–90. https://doi.org/10.1002/aur.1966.
Lv, M.N., Zhang, H., Shu, Y., Chen, S., Hu, Y.Y., and Zhou, M., "The Neonatal Levels of TWG, NW# and CK-BB in Autism Spectrum Disorder from Southern China," *Translational Neuroscience* 7, no. 1 (2016): 6–11. https://doi.org/10.1515/tnsci-2016-0002.
Stancioiu, F., Bogdan, R., and Dumitrescu, R., "Neuron-Specific Enolase (NSE) as a Biomarker for Autistic Spectrum Disease (ASD)," *Life (Basel)* 13, no. 8 (2023). https://doi.org/10.3390/life13081736.

431 Ham, A., Chang, A.Y., Li, H., et al., "Impaired Macroautophagy Confers Substantial Risk for Intellectual Disability in Children with Autism Spectrum Disorders," *Molecular Psychiatry* (2024). https://doi.org/10.1038/s41380-024-02741-z.

432 Glick, D., Barth, S., and Macleod, K.F., "Autophagy: Cellular and Molecular Mechanisms," *Journal of Pathology* 221, no. 1 (2010): 3–12. https://doi.org/10.1002/path.2697.

433 Du, K., Farhood, A., and Jaeschke, H., "Mitochondria-Targeted Antioxidant Mito-Tempo Protects against Acetaminophen Hepatotoxicity," *Archives of Toxicology* 91, no. 2 (2017): 761–73. https://doi.org/10.1007/s00204-016-1692-0.

434 Cendejas-Hernandez, J., Sarafian, J., Lawton, V., et al., "Paracetamol (Acetaminophen) Use in Infants and Children Was Never Shown to Be Safe for Neurodevelopment: A Systematic Review with Citation Tracking," *European Journal of Pediatrics* 181 (2022): 1835–57. https://doi.org/10.1007/s00431-022-04407-w.

435 Cendejas-Hernandez, J., Sarafian, J., Lawton, V., et al., "Paracetamol (Acetaminophen) Use in Infants and Children Was Never Shown to Be Safe for Neurodevelopment: A Systematic Review with Citation Tracking," *European Journal of Pediatrics* 181 (2022): 1835–57. https://doi.org/10.1007/s00431-022-04407-w.

436 Kinoshita, M., Stempel, K.S., Borges do Nascimento, I.J., and Bruschettini, M., "Systemic Opioids Versus Other Analgesics and Sedatives for Postoperative Pain in Neonates," *Cochrane Database of Systematic Reviews* 3, no. 3 (2023): Cd014876. https://doi.org/10.1002/14651858.CD014876.pub2.

437 Skovlund, E., Handal, M., Selmer, R., Brandlistuen, R.E., and Skurtveit, S., "Language Competence and Communication Skills in 3-Year-Old Children after Prenatal Exposure to Analgesic Opioids," *Pharmacoepidemiology and Drug Safety* 26, no. 6 (2017): 625–34. https://doi.org/10.1002/pds.4170.

438 Labba, N.-A., Wæhler, H.A., Houdaifi, N., et al., "Paracetamol Perturbs Neuronal Arborization and Disrupts the Cytoskeletal Proteins SPTBN1 and TUBB3 in Both Human and Chicken in Vitro Models," *Toxicology and Applied Pharmacology* 449 (2022): 116130. https://doi.org/10.1016/j.taap.2022.116130.

439 Posadas, I., Santos, P., Blanco, A., Muñoz-Fernández, M., and Ceña, V., "Acetaminophen Induces Apoptosis in Rat Cortical Neurons," *PLOS One* 5, no. 12 (2010): e15360. https://doi.org/10.1371/journal.pone.0015360.
Vigo, M.B., Pérez, M.J., De Fino, F., et al., "Acute Acetaminophen Intoxication Induces Direct Neurotoxicity in Rats Manifested as Astrogliosis and Decreased Dopaminergic Markers in Brain Areas Associated with Locomotor Regulation," *Biochemical Pharmacology* 170 (2019): 113662. https://doi.org/https://doi.org/10.1016/j.bcp.2019.113662.

440 Schultz, S., DeSilva, M., Gu, T.T., Qiang, M., and Whang, K., "Effects of the Analgesic Acetaminophen (Paracetamol) and Its Para-Aminophenol Metabolite on Viability of Mouse-Cultured Cortical Neurons," *Basic & Clinical Pharmacology & Toxicology* 110, no. 2 (2012): 141–44. https://doi.org/10.1111/j.1742-7843.2011.00767.x.

441 Labba, N.-A., Wæhler, H.A., Houdaifi, N., et al., "Paracetamol Perturbs Neuronal Arborization and Disrupts the Cytoskeletal Proteins SPTBN1 and TUBB3 in Both Human and Chicken in Vitro Models," *Toxicology and Applied Pharmacology* 449 (2022): 116130. https://doi.org/https://doi.org/10.1016/j.taap.2022.116130.

442 Roberts, I.D., Krajbich, I., and Way, B.M., "Acetaminophen Influences Social and Economic Trust," *Scientific Reports* 9, no. 1 (2019): 4060. https://doi.org/10.1038/s41598-019-40093-9.

443 DeWall, C.N., MacDonald, G., Webster, G.D., et al., "Acetaminophen Reduces Social Pain: Behavioral and Neural Evidence," *Psychological Science* 21, no. 7 (2010): 931–37. https://doi.org/10.1177/0956797610374741.

444 Durso, G.R.O., Luttrell, A., and Way, B.M., "Over-the-Counter Relief from Pains and Pleasures Alike: Acetaminophen Blunts Evaluation Sensitivity to Both Negative and Positive Stimuli," *Psychological Science* 26, no. 6 (2015): 750–58. https://doi.org/10.1177/0956797615570366.

445 Randles, D., Kam, J.W.Y., Heine, S.J., Inzlicht, M., and Handy, T.C., "Acetaminophen Attenuates Error Evaluation in Cortex," *Social Cognitive and Affective Neuroscience* 11, no. 6 (2016): 899–906. https://doi.org/10.1093/scan/nsw023.

446 St Omer, V.V., and McKnight, E.D., 3rd., "Acetylcysteine for Treatment of Acetaminophen Toxicosis in the Cat," *Journal of the American Veterinary Medical Association* 176, no. 9 (1980): 911–13.
Court, M.H., "Feline Drug Metabolism and Disposition: Pharmacokinetic Evidence for Species Differences and Molecular Mechanisms," *Veterinary Clinics of North America: Small Animal Practices* 43, no. 5 (2013): 1039–54. https://doi.org/10.1016/j.cvsm.2013.05.002.
Anvik, J.O., "Acetaminophen Toxicosis in a Cat," *Canadian Veterinary Journal* 25, no. 12 (1984): 445–47.
Savides, M.C., Oehme, F.W., Nash, S.L., and Leipold, H.W., "The Toxicity and Biotransformation of Single Doses of Acetaminophen in Dogs and Cats," *Toxicology and Applied Pharmacology* 74, no. 1 (1984): 26–34. https://doi.org/10.1016/0041-008x(84)90266-7.
Lautz, L.S., Jeddi, M.Z., Girolami, F., Nebbia, C., and Dorne, J., "Metabolism and Pharmacokinetics of Pharmaceuticals in Cats (Felix Sylvestris Catus) and Implications for the Risk Assessment of Feed Additives and Contaminants," *Toxicology Letters* 338 (2021): 114–27. https://doi.org/10.1016/j.toxlet.2020.11.014.

447 Miller, R.P., Roberts, R.J., and Fischer, L.J., "Acetaminophen Elimination Kinetics in Neonates, Children, and Adults," *Clinical Pharmacology & Therapeutics* 19, no. 3 (1976): 284–94. https://doi.org/10.1002/cpt1976193284.
Cook, S.F., Stockmann, C., Samiee-Zafarghandy, S., et al., "Neonatal Maturation of Paracetamol (Acetaminophen) Glucuronidation, Sulfation, and Oxidation Based on a Parent-Metabolite Population Pharmacokinetic Model," *Clinical Pharmacokinetics* 55, no. 11 (2016): 1395–411. https://doi.org/10.1007/s40262-016-0408-1.

448 Green, M.D., Shires, T.K., and Fischer, L.J., "Hepatotoxicity of Acetaminophen in Neonatal and Young Rats. I. Age-Related Changes in Susceptibility," *Toxicology and Applied Pharmacology* 74, no. 1 (1984): 116–24. https://doi.org/10.1016/0041-008x(84)90277-1.

449 Viberg, H., Eriksson, P., Gordh, T., and Fredriksson, A., "Paracetamol (Acetaminophen) Administration During Neonatal Brain Development Affects Cognitive Function and Alters Its Analgesic and Anxiolytic Response in Adult Male Mice," *Toxicological Sciences* 138, no. 1 (2013): 139–47. https://doi.org/10.1093/toxsci/kft329.

450 Freed, G.L., Clark, S.J., Butchart, A.T., Singer, D.C., and Davis, M.M., "Parental Vaccine Safety Concerns in 2009," *Pediatrics* 125, no. 4 (2010): 654–59. https://doi.org/10.1542/peds.2009-1962.
Bazzano, A., Zeldin, A., Schuster, E., Barrett, C., and Lehrer, D., "Vaccine-Related Beliefs and Practices of Parents of Children with Autism Spectrum Disorders," *American Journal on Intellectual and Developmental Disabilities* 117, no. 3 (2012): 233–42. https://doi.org/10.1352/1944-7558-117.3.233.

451 Wakefield, A.J., Murch, S.H., Anthony, A., et al., "Ileal-Lymphoid-Nodular Hyperplasia, Non-Specific Colitis, and Pervasive Developmental Disorder in Children," *Lancet* 351, no. 9103 (1998): 637–41. https://doi.org/10.1016/s0140-6736(97)11096-0.

452 Schultz, S.T., Klonoff-Cohen, H.S., Wingard, D.L., Akshoomoff, N.A., Macera, C.A., and Ji, M., "Acetaminophen (Paracetamol) Use, Measles-Mumps-Rubella Vaccination, and Autistic Disorder. The Results of a Parent Survey," *Autism* 12, no. 3 (2008): 293–307.
Schultz, S., *Understanding Autism: My Quest for Nathan* (Schultz Publishing LLC, 2013).

453 Maher, B., "Personal Genomes: The Case of the Missing Heritability," *Nature* 456, no. 7218 (2008): 18–21. https://doi.org/10.1038/456018a.

454 Maher, B., "Personal Genomes: The Case of the Missing Heritability," *Nature* 456, no. 7218 (2008): 18–21. https://doi.org/10.1038/456018a.

455 Roberts, A.L., Lyall, K., Rich-Edwards, J.W., Ascherio, A., and Weisskopf, M.G., "Maternal Exposure to Childhood Abuse Is Associated with Elevated Risk of Autism," *JAMA Psychiatry* 70, no. 5 (2013): 508–15. https://doi.org/10.1001/jamapsychiatry.2013.447.

456 Jones, J.P., 3rd, Williamson, L., Konsoula, Z., Anderson, R., Reissner, K.J., and Parker, W., "Evaluating the Role of Susceptibility Inducing Cofactors and of Acetaminophen in the Etiology of Autism Spectrum Disorder," *Life (Basel)* 14, no. 8 (2024). https://doi.org/10.3390/life14080918.

457 Clayton-Smith, J., Bromley, R., Dean, J., et al., "Diagnosis and Management of Individuals with Fetal Valproate Spectrum Disorder; a Consensus Statement from the European Reference Network for Congenital Malformations and Intellectual Disability," *Orphanet Journal of Rare Diseases* 14, no. 1 (2019): 180. https://doi.org/10.1186/s13023-019-1064-y.

458 Zhao, L., Jones, J., Anderson, L., et al., "Acetaminophen Causes Neurodevelopmental Injury in Susceptible Babies and Children: No Valid Rationale for Controversy," *Clinical and Experimental Pediatrics* (2023). https://doi.org/10.3345/cep.2022.01319.
Jones, J.P., 3rd, Williamson, L., Konsoula, Z., Anderson, R., Reissner, K.J., and Parker, W., "Evaluating the Role of Susceptibility Inducing Cofactors and of Acetaminophen in the Etiology of Autism Spectrum Disorder," *Life (Basel)* 14, no. 8 (2024). https://doi.org/10.3390/life14080918.

GLOSSARY OF TERMS AND NOTED INDIVIDUALS

Acetaminophen (paracetamol; N-acetyl-p-aminophenol; APAP): A neurodevelopmental toxin widely used as a drug to treat fevers and pain. Acetaminophen induces autism spectrum disorder in susceptible individuals from birth to about age six, and probably during pregnancy as well. It is often sold under the tradename Tylenol in the US, whereas Calpol and Panadol are popular tradenames in Europe. It is known by over 100 other names, depending on country of sale, including Aeknil, Caffetin, Paralen, Paramax, Paramol, Perfalgan, Saridon, Tachipirina, Tafirol, Tempra, Termalgin, and Ultracet.

Acetanilide (N-phenylacetamide): A pain reliever and fever reducer introduced for commercial use in 1886. Acetanilide is converted into acetaminophen by enzymes present in the human body (Chapter 2). By the 1970s, acetanilide had fallen into disuse due to well-known toxic side effects. In 1983, the drug was banned in the US due to risks of blood, liver, and kidney damage.

Ad hominem fallacy: Attacks on the character, motivation, or personal traits of the messenger in an attempt to discredit an argument (Chapter 9).

Ahlqvist, Viktor: A scientist and the first author on a highly influential study on acetaminophen use in pregnancy using data from Sweden and published in the *Journal of the American Medical Association* in 2024. Ahlqvist worked with Brian K. Lee at Drexel University, the senior author on the study. They unintentionally demonstrated that adjusting for factors related to oxidative stress—including genetics—eliminates apparent associations between acetaminophen use during pregnancy and autism. The study supported dozens of other lines of evidence indicating that acetaminophen induces autism only in the presence of oxidative stress, but it was—unfortunately, tragically, and widely—misinterpreted as indicating that acetaminophen is safe during pregnancy (Introduction; Chapter 3; Chapter 5; Chapter 10/ 2x).

Appeal to authority fallacy: The argument that something is true because an "expert" or respected organization says it is true (Chapter 9). Paradigm shifts occur when experts and respected organizations are wrong. See also Planck, Max, The Planck Principle, and Semmelweis, Ignaz.

Asperger, Hans: An Austrian physician and pediatrician who described a subset of patients with autism in 1944. After his seminal work describing autism, he was appointed Chair of Pediatrics at the University of Vienna. The form of autism he described was later briefly named *Asperger syndrome* until complaints arose over his cooperation with the Nazi occupation (Chapter 2).

Associations between acetaminophen use and autism through time: Refers to correlations between changes in the use of acetaminophen (or acetanilide or phenacetin) over the past 140 years and changes in the prevalence of autism (Chapter 2 / 3x; Chapter 4 throughout).

Associations between acetaminophen and autism with place: Refers to correlations between differences in acetaminophen use in different parts of the world and differences in the prevalence of autism (Chapter 4/ 2x) Chapter 4 Table 4.1 (Chapter 4).

Associations between circumcision and autism: One study of the Danish population by Frisch and Simonsen in 2015 found approximately two-fold more infantile autism in circumcised boys than uncircumcised boys. Acetaminophen is often used by physicians and by parents to reduce pain from the procedure. The idea that acetaminophen use is riskiest shortly after birth is supported by this study and several other lines of evidence (Chapter 7).

The Atlantic: A prestigious media outlet, generally considered to lean left politically, that published an article on September 9, 2025, misrepresenting evidence of a causal relationship between acetaminophen and autism (Chapter 1).

Banana, Joe: A fictitious character used by the author (Parker) for the purpose of illustrating aberrant human behavior (Chapter 3; Chapter 7). Mr. Banana is acquainted with Amy Tomato, who is taller than the average female. Amy Tomato and Nancy Cucumber, mentioned in Chapters 7 and 11, respectively, are also fictitious.

Bollinger, Ralph Randal (Randy): An immunologist and transplant surgeon best known for co-discovering the function of the human vermiform appendix with the author (Parker). He worked at Duke University for his entire career, serving as Chief of General Surgery for much of that time. He has worked with the author (Parker) for more than three decades (forward; Chapter 1/ 2x; Chapter 4).

Cherry-picking fallacy (fallacy of incomplete evidence): Embracing evidence that supports a particular conclusion while ignoring contradictory evidence. In scientific practice, cherry picking subverts the progress of science and is widely considered unethical. It can occur within a laboratory (Chapter 8), but editorial selection of articles supporting particular conclusions, described in Chapter 3 and 7, is cherry picking on a grander scale. See also *Pediatrics*, the journal, and *Environmental Health*, the journal.

Citation tracking (citation analysis or tracing): A method of following published work forward or backward in time using the references or citations listed in the scientific literature. The method is often used to identify which studies are most influential but can also be used to establish the basis for assertions made in the literature (Chapter 7).

Clinical and Experimental Pediatrics: The flagship journal of the Korean Pediatric Society. The journal published a paper in 2023 that concluded the risks of using acetaminophen from birth to age six outweigh the benefits (Chapter 2).

Codeine: A drug that is converted by the human body into morphine, a potent narcotic pain reliever. The drug itself has no pain-relieving effect prior to its conversion into morphine. The enzyme CYP2D6 converts codeine to morphine, making codeine dangerous for individuals with exceptionally high levels of CYP2D6 activity. The drug was introduced in the 1830s for pain relief. After about 170 years of use, in the early 2010s, its use in children was restricted due to its tendency to kill those who are susceptible (Chapter 3).

Conclusions regarding the role of acetaminophen in the of autism: Conclusions are listed three times in the book (introduction; Chapter 2; Chapter 10).

Confirmation bias fallacy: Seeking information or interpreting information in a way that supports a particular view or conclusion. Individuals with higher intelligence can be more susceptible to this error, which is related to, but distinct from, the cherry-picking fallacy (Chapter 9).

Confounding factors: Factors that influence the connection between *other* factors in a way that is potentially misleading. In general, confounding factors *cause* an outcome but are linked to another, non-causal factor, making it appear as if the non-causal factor is the cause (Chapter 3 throughout).

Consensus bias fallacy: The delusion that most others share similar beliefs. In terms of the connection between acetaminophen and autism, it is thinking that all *reasonable* people share similar beliefs, and that others must be unreasonable or irrational (Chapter 9).

Cord blood: The blood present in the umbilical cord. Cord blood is part of the fetal circulation and, therefore, has the same properties as fetal blood. When collected at the time of birth, it can be used to determine what was in the newborn's blood at the moment of birth. Yuelong Ji, working with Xiaobin Wang at Johns Hopkins University, found strong associations between acetaminophen metabolites in cord blood and autism (Chapter 2; Chapter 5).

Destroy the Weak Link fallacy: Attempts to discredit scientific conclusions by attacking a subsection of evidence rather than looking at the whole body of evidence. The approach is based on the proverb that "a chain is only as strong as its weakest link"; however, scientific conclusions are often supported by dozens of lines of evidence, some stronger than others. In these cases, a more applicable proverb is that "a cord of many strands is not easily broken" (Chapter 8).

Emotional compromise: The state in which logical reasoning is impaired or subverted by emotions triggered by facts or observations (Chapter 10).

***Environmental Health*, the journal**: A scientific journal which published a paper from Andrea Baccarelli's lab in 2025 claiming—with no scientific basis and contrary to all scientific evidence—that the interaction between oxidative stress and acetaminophen is "speculative." The journal editors refused to publish scientific evidence demonstrating that Baccarelli's claim was verifiably false. The Baccarelli group's error undermines the analysis they performed, resulting in a profound underestimation of the impact of acetaminophen on neurodevelopment during pregnancy (Chapter 3; Chapter 10). See also Prada, Diddier.

Error of Infinite Investigation: This error involves asking for more evidence and more research before a conclusion can be reached, regardless of how much evidence already exists. This approach focuses on the *unknowns* rather than the *knowns*, which can prevent implementation of urgently needed public health policies (Chapter 5; Chapter 8).

Evidence connecting acetaminophen and autism, available in 2008: Scientific evidence available in 2008 and 2009 connecting acetaminophen with the induction of autism was quite strong (Chapter 2).

Evidence connecting acetaminophen and autism, summary of all: The approximately 30 lines of evidence available in early 2025 pointing toward the induction of autism through exposure of susceptible children to acetaminophen (Chapter 2, Tables 1 through 4).

Fetal alcohol spectrum disorder (FASD): A group of conditions caused by prenatal exposure to alcohol. FASD results in a range of physical, mental, and behavioral effects, depending on the affected individual. FASD is distinct from autism spectrum disorder (ASD) but shares many of its features (Chapter 7).

Glucuronidation: An important metabolic pathway that leads to safe (non-toxic) metabolism of acetaminophen. This pathway is impaired in newborns and in individuals with autism (Chapter 7). This pathway is also deficient in domestic cats, resulting in unacceptable levels of toxicity from acetaminophen exposure (Chapter 7).

Glutathione: The body's master antioxidant, necessary to detoxify NAPQI, the metabolite of acetaminophen produced by oxidation. The amount of available (active) glutathione is often used by scientists as a measure of oxidative stress (Chapter 3). See also NAPQI.

Good, Peter: An independent scientist, the first to take a broad perspective on the connection between acetaminophen and autism, collecting evidence from multiple lines of investigation and concluding that acetaminophen use could be responsible for the increased prevalence of autism. Good's first work on the topic was published in 2009 in the journal *Alternative Medicine Review* (Chapter 1/ 2x; Chapter 8).

h-index: One measure of a scientist's productivity. The h-index is defined as the largest number of papers published by a scientist that has received at least that number of citations. For example, a scientist with exactly 20 papers that have each been cited at least 20 times would have an h-index of 20 (Chapter 1).

Illusion of Consensus (a false consensus): A "consensus" derived from multiple sources repeating conclusions that come from a single source rather than drawing those conclusions independently (Introduction).

Infantile autism: A subset of autism in which the condition appears to be present from infancy. In these cases, no regression or loss of developmental skills is observed (Chapter 2).

Interacting variables: Factors or variables that, when combined, cause an effect. The example seen throughout the book is that acetaminophen interreacts with a wide range of factors related to oxidative stress to induce autism (Chapter 3, throughout).

Invincible Ignorance Fallacy: A logical error in which no amount of evidence, regardless of how compelling, is viewed as convincing (Chapter 5; Chapter 9). See also confirmation bias fallacy.

Jones, John P., III: A clinical pharmacist and first author on a paper published in *Life* in 2024 demonstrating that the data published by Ahlqvist et al. in 2024 are exactly what would be predicted if acetaminophen combined with oxidative stress causes autism (Chapter 3; Chapter 10). Jones has collaborated with the author (Parker) for three decades. See also Ahlqvist, Viktor.

Journal of Xenobiotics: A journal for studies related to all aspects of xenobiotics. Xenobiotics are substances not normally produced by the organism in which they are found; examples include drugs, pollutants, and food additives. Based on a peer review exhibiting obvious and extensive signs of bias (Appendix A), the journal's editors rejected a manuscript concluding that acetaminophen induces autism in susceptible individuals (Introduction).

Kanner, Chaskel Leib (Leo): A pediatric psychiatrist at Johns Hopkins Hospital who described a subset of patients with autism in a 1943 paper (Chapter 2/ 2x).

Kekatos, Mary: A reporter for ABC News who wrote an article "debunking" the dangers of taking acetaminophen during neurodevelopment (Chapter 4; Chapter 11).

Kennedy, Robert Francis Jr. (RFK Jr.; Bobby): Secretary of Health and Human Services during the second term of US president Donald Trump. Kennedy brought awareness of the connection between acetaminophen and autism to the Trump administration (Chapter 1).

Klotman, Mary Frances Earley: A clinician and expert in some aspects of HIV infection who was dean of the School of Medicine at Duke University starting in 2017. She has been married to Paul Klotman since 1981 (Chapter 1). See also Klotman, Paul.

Klotman, Paul: A nephrologist and expert in some aspects of HIV infection, Klotman has publicly stated that litigating the idea that acetaminophen induces autism during pregnancy is "just silly" (Introduction; Chapter 1). He has been the president, CEO, and executive dean of the Baylor College of Medicine since 2010 and has been married to Mary Klotman since 1981. See also Klotman, Mary Frances Earley.

Meza-Keuthen, Maria Susanne: A scientist with a master's degree in chemistry and a public-school counselor with 25 years of counseling experience, she has been married to the author since 1989.

Minerva Pediatrics: The oldest pediatric journal in Italy. In 2022, the journal published the first extensive review concluding "without any reasonable doubt that oxidative stress puts some babies and children at risk of paracetamol-induced neurodevelopmental injury, and that postnatal exposure to paracetamol in those susceptible babies and children is responsible for many if not most cases of ASD." (Chapter 2)

NAPQI (N-acetyl-p-quinoneimine): The toxic metabolite of acetaminophen. Glutathione is required to safely neutralize NAPQI. Conditions present in children with autism tend to favor formation of NAPQI and disfavor its safe neutralization (Chapter 2; Chapter 3).

Noble Disobedience: The widely accepted principle that rules should be broken or disregarded when the consequences of following those rules are immoral. Unfortunately, disregard for the scientific process, regardless of how well intentioned, does not result in laudable outcomes (Introduction).

Non Sequitur Fallacy: A statement or conclusion that does not logically follow from the justification given (Chapter 9).

Occam's Razor: The principle that the simplest explanation of the observations is likely the correct explanation. A logical consequence of the Razor is that the magnitude of the difference in complexity between two explanations, both consistent with the observations, is proportional to the likelihood that the simpler explanation is correct (Chapter 3; Chapter 4; Chapter 8; Chapter 9).

Oxidative Stress: The condition in which the body cannot keep up with the demands placed on it, leading to a buildup of harmful oxidants and a deficiency of antioxidants. A wide range of genetic, epigenetic, and environmental factors can lead to oxidative stress. The conclusion that acetaminophen interacts with oxidative stress to induce autism is considered in detail in Chapter 3.

Patel, Alok: A pediatrician at Stanford University, Patel has publicly criticized the conclusion that acetaminophen exposure during pregnancy induces autism in susceptible offspring (Chapter 11).

***Pediatrics*, the journal**: The highly influential flagship journal for the American Academy of Pediatrics. The journal published some of the seminal work that encouraged acetaminophen use in babies and children and has blocked publication of overwhelming evidence demonstrating that it is not safe for neurodevelopment (Chapter 3; Chapter 5/ 2x; appendix B; appendix C; Chapter 7/ 2x; Chapter 8; Chapter 9).

Peer-review process: The assessment of a scientist's proposals or work by other scientists. The process is used to judge which projects should be funded and which science should be published (Introduction). The importance of timely review and the vital role of journal editors is described in Chapters 4 and 5, respectively. Approaches to avoid offending editors are described in Chapter 7. The peer-review process combined with insufficient funding for scientists results in particularly adverse effects. (See also threat rigidity.) Exclusive review processes effectively block new errors from being published and are considered the most rigorous with the highest degree of academic and ethical integrity. However, this same exclusiveness also leads to entrenched biases that make it difficult to identify even obvious errors. See also Planck, Max, The Planck Principle, and Semmelweis, Ignaz.

Phenacetin (N-(4-ethoxyphenyl)acetamide): A pain reliever and fever reducer introduced for commercial use in 1887. Like acetanilide, phenacetin is converted into acetaminophen by enzymes present in the human body. In 1983, after almost a century of popular use as a drug, it was removed from the consumer market in the US due to risks of cancer and kidney damage (Chapter 2).

The Planck Principle: The assertion that science moves forward when entrenched, older scientists retire and are replaced by younger scientists able to embrace new ideas. The principle was described in 1950 by Max Planck, the originator of quantum theory and one of the founders of modern physics (Introduction; Chapter 9).

Prada, Diddier: An epidemiologist at the Mount Sinai Hospital and the first author of a study2 published in *Environmental Health* in 2025 claiming that oxidative stress may not necessarily affect acetaminophen toxicity. Andrea Baccarelli at Harvard University was the senior author on the study. The study assumed (incorrectly) that acetaminophen does *not* interact with oxidative stress in the analysis of their data, obscuring the relationship between acetaminophen and autism (Introduction; Chapter 3). See also *Environmental Health*, the journal.

Regressive autism: A subset of autism in which some developmental skills, such as speech and gross motor coordination, are lost prior to diagnosis with autism (Chapter 2; Chapter 5).

Retaliatory activity: Punitive actions aimed at an employee in response to a perceived offense. Punishments can include unsubstantiated accusations of poor performance, demotion or loss of opportunity for promotion, decreased pay, loss of access to resources, loss of responsibility, and firing (Chapter 1).

Schultz, Stephen: A dentist with a son who regressed into autism at 13 months of age after receiving the MMR vaccine followed by treatment with acetaminophen every four hours for about three weeks. Schultz was the first to test that idea that acetaminophen use and autism are connected. Schultz earned his PhD conducting a case-controlled study that found regressive autism was strongly associated with acetaminophen use in early childhood (Chapter 1; Chapter 2; Chapter 3; Chapter 5; Chapter 7; Chapter 8; Chapter 10).

Science Bashing: The unethical use of scientific jargon to make invalid accusations designed to discredit a scientific study—to slander it in other words. (Chapter 8)

Scientific Bluff: The unethical presentation of scientific information as fact when it is not supported by the scientific literature (Chapter 8).

Semmelweis, Ignaz: A Hungarian physician and scientist, Semmelweis was one of the first to promote hand washing and the wearing of clean clothes for the purpose of reducing infection during surgery. Semmelweis's treatise on the topic was published in 1861, and the English translation of that document is cited in the Endnotes. Despite overwhelming evidence that Semmelweis's recommendations were effective, the most prestigious scientists and doctors of the day rejected his conclusions (Chapter 3).

Shaw, William (Bill): An independent scientist, the second person to take a broad perspective on the connection between acetaminophen and autism, collecting evidence from multiple lines of investigation and concluding that acetaminophen use could be responsible to the increased prevalence of autism. Shaw's first work on the topic was published in 2013 in the *Journal of Restorative Medicine* (Chapter 1/ 2x; Chapter 8).

Sukhareva, Grunya: A pediatric psychiatrist, Sukhareva was the first person to describe the hallmarks of autism spectrum disorder in the medical literature (Chapter 2/ 2x). Sukhareva's work was published in 1925, and a description of her life and work is cited in the Endnotes.

Sulfation: An important metabolic pathway that leads to safe (non-toxic) metabolism of acetaminophen. This pathway is frequently impaired in individuals with autism (Chapter 7).

Susceptibility to acetaminophen-mediated neurodevelopmental problems: Elevated risks caused by oxidative stress. A variety of genetic, epigenetic, and environmental factors contribute to oxidative stress (Chapter 4/ 2x). Susceptibility is discussed as a function of age in Chapter 5 and illustrated in Figures 5.1 and 5.2.

Threat rigidity: A psychological and organizational response to a threat, real or perceived, characterized by inflexibility, tunnel vision, adherence to familiar, established behaviors, and rejection of new information. Threat rigidity stifles innovation and drives conformity. The combination of peer review and limited funding creates enough uncertainty to create threat rigidity (Chapter 9).

Trump, Donald: The 45th and 47th US president who, on September 22, 2025, announced at a White House press conference that acetaminophen should be avoided during pregnancy and after birth (Introduction; Chapter 1).

Vaccination, association with autism: A medical procedure that frequently causes fever and/or pain—usually treated with acetaminophen. This can induce autism in susceptible children (Chapter 5/ 2 case reports; Chapter 8).

Viberg, Henrik: A scientist at Uppsala University and the first individual to test the effects of acetaminophen on neurodevelopment in laboratory animals. Viberg's first work on the topic was published in 2014 (early online in 2013) in the journal *Toxicological Sciences*. That study showed profound impairment of brain development even at weight-adjusted doses *below* those administered to some babies (Chapter 6; Chapter 10).